No. 505
$7.95

INSTALLING

&

SERVICING

HOME AUDIO SYSTEMS

By Jack Hobbs

TAB BOOKS
BLUE RIDGE SUMMIT, PA. 17214

FIRST EDITION

FIRST PRINTING — OCTOBER 1969

Copyright © 1969 by TAB BOOKS

Printed in the United States
of America

Library of Congress Card Number: 75-94452

Preface

This book can be worth its weight in gold—if you read, study, and use the material contained in it. It is not intended for hobbyists, or amateurs, nor does it make any attempt to "play-up" to the so-called professional who, through no fault of his own, has developed a "pin-point-me-the-trouble, do-it-yourself complex"—a complication that grew primarily out of misdirected early technical training.

The book is dedicated exclusively to those who "think": the sincere professionals who have a desire to excel and "move up"; determined service dealers and technicians who make their living in part as diversified operators or wholly as specialists selling, installing, and servicing home, business, educational and industrial audio- and audio-visual equipment; ambitious apprentices and technical-school students who would like to "get-ahead" in the audio area of the electronics field. The book is particularly directed to those now employed as home-entertainment equipment technicians who would like to enhance their knowledge so they can make more money and enjoy better living.

It is assumed here that the reader has either an adequate basic electronics education or, in the case of apprentices and technical-school students, further education is now being obtained through formal technical training and working experience. Very little elementary theory will be repeated here, and that only when necessary to firmly establish concepts which are closely related to everyday, practical problems.

We will cover the practical aspects of installing and servicing all kinds of audio communications equipment used in the home, business, and industry. Every effort has been made here to provide the latest, most accurate, up-to-date, un‑rehashed technical information and state‑of‑the‑art troubleshooting and repair techniques. You will not find, however, such things as "do-it-yourselfer"—type symptom-and-cause charts, alignment procedures, and other information already in the service-data files of all service-dealers and readily available to working technicians and apprentices How well the author has succeeded in his efforts can best be judged only by you—the reader.

Duluth, Minn. 1969

Jack Hobbs
Former Managing Editor,
ELECTRONIC TECHNICIAN/DEALER
Magazine

CONTENTS

1 The Audio Business—Opportunity Unlimited

Getting Your Share — The Modus Operandi—Our Work Methods — Circuit Theory — Your Working Tools   **7**

2 Test Instruments For Audio

The AC Voltmeter — Sine/Square-Wave Generator — IM and Harmonic Distortion Measurements — Harmonic Distortion Meter — Wave Analyzer — Flutter Meter — Noise Level Meter   **12**

3 AM Receiver Maintenance

Basic Troubleshooting Techniques — Dead Radio — Signal Injection — Other Approaches — Converter, Oscillator/Mixer Problems — Troubleshooting Converters — "Tough Dog" Type Problems — Substitute Oscillator — Checking Oscillator/Mixer Stages — Weak Or Weak and Distorted Reception— Distortion—Normal Volume — Noisy Radios — Humming and Motorboating — Intermittents — Isolating Intermittent Faults   **19**

4 Working With FM Tuners

FM Tuner Principles — FM Detectors — Discriminators vs Ratio Detector — Automatic Frequency Control — Varicap AFC Circuits — Limiter Circuits — Troubleshooting FM Tuners — The Oscillator or Converter — Checking For Drift Causes — Regeneration — Solving Regeneration Problems — Inadequate FM Limiting—Noisy Output — Tuner Sensitivity — Distorted Output   **44**

5 Servicing FM Stereo Equipment

FM Multiplex — FM Multiplex Receivers — Mute Gate — Balanced Multiplex Detector — Multiplex Alignments — Channel-Separation Checks — db Separation — FM Receiver Sensitivity — Decoder Circuit Variations   **61**

6 Preamps And Amplifiers

Grounded-Emitter Amplifiers — The Grounded-Collector Amplifier — Typical Preamp — Equalizing Networks — Feedback — Negative Feedback — Current and Voltage Feedback — Bootstrapping — Positive Feedback — Power Amplifiers — Single-

Ended Class B — Modern Hi Fi Stereo Amplifiers — Stereo Amplifier Arrangements — Troubleshooting Modern Audio Amplifiers — General Service Procedure — Hum and Interference Problems — Manufacturer's Service Aids — Integrated Circuits 82

7 'Combinations' And Related Problems
Simple Combination — Combination Switching — Working With Combinations 118

8 Solid-State Power Supplies
Halfwave Rectifiers — Split-Voltage Arrangements — Voltage-Regulating Elements — Troubleshooting 125

9 Tape Recorders And Players
A "Typical" Tape Recorder — Tape Transport — Erase and R/P Heads — Bias/Erase Oscillator — General Circuit Variations — Mechanical Trouble Symptoms — General Mechanical Maintenance — Head Wear, Care, and Adjustment — Head and Tape Guide Degaussing — Height and Azimuth Adjustments — Checking Record Bias and Erase Head Current — Cartridge Recorders — Tape Transport — Viking Cartridge Recorder — Norelco "Cassette" — Home Cartridge Players — "Typical" 8-Track Stereo Deck 138

10 Mobile Radios And Tape Players
General Considerations — Service Literature — Servicing Mobile Radios — Troubleshooting and Precautions — Noise Suppression — Reverberation Equipment — Mobile Tape-Players — Maintenance and Installation Problems 167

11 Servicing Automatic Record Players
Cycling — Cycling Problems — Tone Arm — General Service Considerations — Cleaners and Lubricants — Styli 188

12 Selling And Installing Audio Systems
Sales/Service Approach — The Home "Listening" Room — Speakers and Headphones — General Considerations 210

13 Commercial Audio Systems
Public Address Equipment — Preliminary Considerations — Available Amplifier Types — Distribution Principles — Basic Speaker Types — Speaker Characteristics — Line Matching — Constant-Voltage Lines — Microphones 219

14 Home And Business Intercoms
Typical Intercom Circuits — Wired Systems — RF System — Other Problems 242

Index 251

CHAPTER 1
The Audio Business— Opportunity Unlimited

The home-and commercial-audio equipment business has become a billion-dollar gold mine in recent times. Sales of home AM and AM/FM radios, hifi stereo equipment, phonographs, tape recorders and decks, home and auto tape players and radios continue to increase. Also, public address systems, intercoms, and other audio and audio-visual equipment sales to business and industrial establishments, to schools, churches and hospitals are constantly on the upswing.

GETTING YOUR SHARE

Up until about 10 years ago, a relatively few, widely-scattered service specialists and service-dealers sold, installed, and did most of the maintenance work on high-class home, business, and industrial audio equipment. A majority of the so-called "independent" TV-radio service-dealers and technicians in this country—while potentially better qualified to handle the business than any other group—avoided or failed to promote sales and service. As a result, much of the audio sales business was lost to distributor/retail chains, department stores, and mail-order houses. Even licensed electrical contractors cut heavily into the business, particularly in commercial and industrial audio equipment areas.

But this situation has been changing fast in recent times, especially in suburban and rural locations. Whether you are now an audio specialist—home, commercial or industrial; whether audio-diversified as a TV-radio service-dealer or plan to go into the business; whether working as an apprentice or going to technical school, you can benefit from the growth now taking place in the audio areas of the electronics field only if you prepare to take advantage of this upsurge. For the established service-dealer, this means taking immediate steps to aggressively promote equipment sales and organize your business so you can provide proper service to a fast-growing, affluent, more discriminating public.

Of course, it will be necessary to concentrate more heavily on the fine points of business management so you will be equipped to solve the problems confronting you in coming years, especially in the area of providing high-class service and product promotion. If you are now working as a TV-radio technician, an apprentice, or if you are a consumer-electronics technician-student in a technical school, you can move upward into a higher income area only if you know all about the equipment involved in audio communications—how it works, how to diagnose faults, troubleshoot and repair it quickly and efficiently. If you feel you've missed the boat, forget it. Tomorrow's "audio gold" supply will be far greater than it has ever been in the past and it will belong exclusively to those who know how to dig for it.

But we need a plan—a method of work...

THE MODUS OPERANDI — OUR WORK METHODS

The overall approach outlined here is not new nor is it difficult to master. A few highly successful technicians, many of whom have become successful service-dealers, have used these methods, in one form or another, for years. The first principles involved in becoming a good technician concern your mental approach—your mental attitude. And the greatest stumbling block in your path will probably be that old straw-ghost: "Theory versus practice."

You may be familiar with the expression, "it may be true in theory, but it doesn't work in practice." Or to paraphrase it in one of many ad nauseam variations, "Theory and practice do not always agree." This is not, as it may appear, a paradox. It is a scientific impossibility. But belief in this aphorism has stopped many potentially excellent technicians dead in their tracks. They've convinced themselves that they are "practical" and "logical" technicians and can get along without theory. Let's get tuned in on our times.

Perhaps the confusion began when the word "theory" was first confused with the word "hypothesis." It shouldn't have, though, because a theory is one thing and a hypothesis is something else. A theory, like the electron theory, for example, is a system to be followed in arriving at certain desired results—like a roadmap which tells you how to get from here to there. Like all scientific theories, the electron theory grew out of controlled scientific research, development, trial-and-error—and practical application. If a theory does not work in practice, then it is a false theory. And every theory that is not false continues to develop to a higher level and becomes refined through practice. Theory and practice exist together like the links in a chain, like the components in a "bootstrap" or a feedback control system—fully interrelated, interdependent, inseparable. So, if we know our electron theory it will always work for us in practice—if the theory is applied correctly.

The next prerequisite in our modus operandi, our method of work, also comes under the "mental attitude" heading. We must learn, for example, to see every comparable piece of equipment—every radio, every AM/FM tuner, every amplifier, every tape recorder—as a "whole thing" composed of a few simple circuits or a few mechanical arrangements, each circuit or mechanical arrangement in turn made up of a few parts.

Have you ever observed a skilled bench technician, assisted by a trained parts-replacer, repairing radios? We observed one a few years ago who handled 75 radios in an 8-hour working day. This technician picked up one radio after another, from pocket- to tabletop-size, all different shapes and sizes. He would look, listen, smell, feel, and then make a few rapid checks, mark one or more components with a grease pencil and place each radio beside the parts replacer. After the parts were replaced by his assistant, the technician would then check each radio to make sure it was working properly. The bench technician's average troubleshooting time was 6.4 minutes per radio—hour after hour.

A skilled bench technician seldom looks at a schematic. Do you know why? Because he has trained himself to see every radio as just another radio. They are all alike to him, all having essentially the same circuits—all specific circuits performing the same basic job. He knows the trouble symptoms associated with each circuit. He knows the RF, the oscillator, mixer (or converter), the detector, the audio driver, the audio output, the AVC in every radio works basically the same way. He has avoided the time-consuming habit of using a schematic as a "crutch"...to "logically" guess where the trouble is located. And we may as well point out here that logic is a system of simple algebra (Boolean) used primarily in digital computer circuits and in one area of philosophical "thought." We are not concerned here with logic circuits nor philosophical thought, but with the science of electronics—the science of diagnosing faults by employing a knowledge of only a small part of the electron theory.

And finally, in the "mental attitude" category, we must avoid thinking of the equipment we work on as being "mechanical." Except for phonographs and tape-transport equipment, our work is predominantly electronic. We do not work on "machines" and we are not "mechanics." Most equipment we work on does not function according to mechanical laws. Hence, to repeat, our overall approach must be electronic, based on the practice of electronic theory, the same way as a good doctor practices the science of medicine.

CIRCUIT THEORY

Our work-method repertory must also include a fundamental knowledge of circuit theory. This does not mean that you must know all functional details of circuit operation as a design engineer must know them. But you must know what function each circuit performs, how the individual components work, and what symptoms arise in the equipment when one or more components become defective.

We must know, for example, how negative (inverse, degenerative) and positive (regenerative) feedback circuits operate and where they are used in various types of audio equipment. How do AFC circuits operate in FM tuners and what is likely to happen if one fails? What's the difference between an FM discriminator-type demodulator and a ratio detector? What's a "mute gate" and stereo indicator circuit and how does it work? What kind of circuits are used to stabilize transistorized amplifiers and what happens if they fail to function properly? What's a bias oscillator in a tape recorder and how do you check it? What's a complementary-symmetry amplifier and how does it operate? In chapters to follow we will briefly but adequately cover these and dozens of other basic circuits and their trouble symptoms.

YOUR WORKING TOOLS

Working tools (test instruments) and how they should be used are included in our modus operandi. And it is your job to study and master test instrument techniques as outlined in manufacturers' operating manuals. Many technicians seldom study these manuals to learn what their test instruments can do. Manufacturers' equipment service data is another working tool. But this material is used only when needed—you do not use it for a crutch to lean on. And do not expect this material to substitute for a lack of troubleshooting knowledge. It won't!

This, then, is a broad, overall but specific view or our modus operandi—our work plan. It has evolved from the interdependent nature of theory and practice and has been tested by thousands of successful technicians, many of whom are successful service-dealers, during the past two decades. And it also will give you successful results if you master it and follow it through.

Now let's get to the "meat and beans"!

CHAPTER 2

Test Instruments For Audio

Most electronic equipment manufacturers have long known that consistent sales are maintained only when satisfactory service is provided. Some manufacturers who ignored this basic concept in the past are no longer in business. And the service-dealer who fails to provide top-grade service for the equipment he sells also will fade away—like the proverbial "old soldier." You cannot do a professional service job, provide the public with the kind of service it demands today for its audio equipment—home-entertainment, commercial or industrial—without the proper test instruments. Of course, you may be able to "get by" with a minimum instrument setup, but it is not recommended that you try.

HOW MUCH? HOW LITTLE?

The variety and quality of the test instruments you need will depend to some extent on how diversified your equipment-servicing is now. If you are now a TV-radio service-dealer who plans to branch out into the overall home-entertainment and commercial audio equipment business, it is assumed that you now have a 20K/volt VOM, a good general-purpose VTVM, in-circuit beta-type transistor tester, emission-type tube tester, sweep generator, RF signal generator, and scope. But there are two schools of thought as to what specialized instruments are required, and both agree that a high-grade AC/VTVM, a sine/square-wave generator and an FM/stereo multiplex generator are basic, plus a noise- or audio-level meter if you do public-address (PA) and certain other commercial audio work. The "all-out" instrument school also insists that instrumentation be provided for measuring both harmonic and intermodulation distortion, as well as wow and flutter. They also insist on a dual - trace, wide - band, triggered-sweep scope, especially for stereo equipment troubleshooting.

Although our position is that you should have all the test instruments you need and can afford to buy, we have certain reservations about the need for specific test instruments that far exceed the capabilities of most equipment they are to be used to service. During the past few years a third school has "jetted-in" on the audio test instrument scene. Their position is: More tolerance is being built into the equipment we service. And test instruments have now been developed which provide a satisfactory "compromise" between cost and capability. One such instrument, a stereo analyzer, also contains a variety of basic test instruments which can be used for general service work in almost all areas covered by this book.

THE AC VOLTMETER

No service-dealer or technician who is serious about audio work will attempt to maintain this equipment without a good electronic AC voltmeter. It is the basic instrument in all high-class audio-equipment service shops. With it, the sine/square-wave generator and a good scope, you can "makeshift" satisfactorily in many practical jobs without other specialized instruments. And this meter should meet certain minimum specifications.

It must be highly sensitive, because we will be measuring small voltage outputs from certain microphone types, phono pickups (dynamic and magnetic), and tape heads. On some equipment the outputs are only a few millivolts (mv). The meter's most sensitive scale, then, should be from 0 to 3 mv. Some lab type AC/VTVMs have scales from 0 to 1 mv or less. Perhaps the maximum scale should range from 30 to 300v.

Because a considerable amount of engineering is involved in designing and manufacturing a good AC/VTVM, they cost money. To obtain the required stability, for example, a large amount of negative feedback (degenerative) is used in the meter's amplifier stages. Heavy negative feedback, as you know, reduces amplifier gain and thus a number of stages are needed for high sensitivity. The meter should be sufficiently stable so a zero-set control is not required on the front panel. Additionally, the meter's amplifier must be designed so it has a flat frequency response from a few Hertz up to at least 500 kHz at an accuracy of ±3%.

The meter scale should read RMS volts and be calibrated logarithmically to prevent crowding in the left or low-voltage area of the meter scale. A linear decibel (db) scale is also convenient so voltage gain and db can be easily transposed and compared directly. These are the principle requirements of an audio-quality VTVM. Also available now are high-grade, battery-operated, portable solid-state AC voltmeters (TVMs).

SINE / SQUARE - WAVE GENERATOR

The next basic specialized instrument needed is a sine/square-wave generator which also must meet certain minimum specifications. For example, the internal harmonic distortion of this instrument should not exceed 0.1% over a frequency spectrum from about 10 Hz to 20 kHz. Rise time should be 0.5 microseconds (μs) or faster. The instrument must be easily and accurately adjustable from 10 Hz up to about 200 to 300 kHz. Frequency stability should be good. The output impedance should not exceed 600 ohms.

The generator's frequency-selecting circuits must be isolated from, and independent of, the output load, and it should have a relatively flat response over its entire frequency range. This feature, however, is not an essential requirement since, in practice, compensation can be made for output level variations. The flat response is more of a convenience than an essential requirement. And the instrument should have a well-filtered power supply having very low ripple content.

IM AND HARMONIC DISTORTION MEASUREMENTS

Although harmonic- and IM-distortion (wave analyzer) instruments are primarily production tools for checking amplifiers (frequently used to compare amplifiers against a "standard" set of specifications), they are excellent troubleshooting instruments for locating and isolating fault - symptoms that sometimes appear in audio amplifiers. It is difficult to explain what harmonic- and IM-distortion is, technically, unless we go into that area of mathematics covered by Fourier analysis of periodic functions. And if we took the time to do that, our audio troubleshooting swords would not be sharpened all that much for the effort.

Perhaps in a somewhat oversimplified way we can satisfactorily define harmonic distortion as a "change that takes place in a relatively perfect sine wave caused by the inclusion of unwanted harmonics of the fundamental." This change takes place in the amplifier somewhere between its input and output. Harmonic distortion is usually expressed as a percentage-ratio between the fundamental test frequency amplitude at the amplifier's input and the harmonic content amplitude at the amplifier's output. The percentage of harmonic distortion varies with an amplifier's output power; that is, it varies with the amount of power output at any given instant. At normal listening levels in the home (generally from a few hundred milliwatts to a few watts at most), harmonic distortion should not exceed one or two percent and it should be not more than one percent or much less in high-grade home-entertainment equipment when operating at low output levels. In many commercial installations where the power output runs high, this distortion rises to five percent or higher, depending on equipment design.

IM distortion is closely related to harmonic distortion, and as a general rule, where one exists the other is present. We can define it, once again in a somewhat oversimplified way, by saying it is "caused by fundamentals and harmonics, sum and difference frequencies, mixing and modulating each other." Both harmonic and IM distortion problems arise because of nonlinear characteristics of amplifier circuitry.

Distortion arises when a tube or transistor drifts into nonlinear areas of their operating characteristic. This "drift" can be caused by "failure" or changes taking place in various circuit components—including tubes, transistors, capacitors, resistors, and varying temperatures. In addition to the aforementioned conditions, the amount of distortion is determined primarily by the amplifier class, how the amplifier is designed, how it is biased and the methods used to stabilize bias. This is only part of the story, but it won't raise our troubleshooting standards much to dig deeper into the subject at this point.

It should be said, however, that the type of distortion analyzers we can afford to buy will give only close, approximate measurements. Although adequate for ordinary "consumer-type equipment, it must be remembered that these instruments will measure the amount of distortion in an amplifier, plus internal hum, plus whatever distortion may exist in the sine/square-wave generator we use to check the amplifier. The entire instrument chain is only as strong as its weakest link.

HARMONIC DISTORTION METER

Simply, a harmonic-distortion meter contains a very sensitive AC/VTVM, an amplifier, and circuitry (usually a Wien-bridge circuit) to filter, or null-out, the fundamental test frequency injected at the amplifier input. The meter reads the percent of harmonic content, including hum, as previously mentioned. This instrument must be able to measure harmonics up to at least the 3rd harmonic of your sine/square-wave generator's 20-kHz frequency. This means up to at least 60 kHz—preferably higher—with linear response over the entire range from at least 60 Hz.

The instrument, of course, should be designed so it produces little distortion and hum. The fundamental suppressing circuitry should be able to reduce a fundamental frequency used in checking to approximately 80 db down without attenuating the 2nd and 3rd harmonics more than a small amount. And the meter should be sufficiently accurate and sensitive to provide at least a 0.1% distortion reading. This type meter is also known as a "fundamental-suppression analyzer," or "total-harmonic analyzer."

WAVE ANALYZER

As previously mentioned, modern "think-tanks" have turned out a new crop of instrument ideas. Exponents of these ideas hold that IM-distortion measurements are more significant than harmonic-distortion measurements, especially on high-class audio equipment. And they maintain that some moderately-priced test instruments are fully adequate to properly check out most "consumer-grade" equipment. Despite this, we'll make a brief pass at the one instrument traditionally

16

used to make accurate IM-D measurements on home- and commercial-type audio amplifiers.

Although a harmonic-distortion meter measures the total amount of distortion in an amplifier, the IM-distortion meter (or wave analyzer) is designed to measure individual harmonics. It uses a balanced bridge (or other methods) to mix two sine waves of different frequencies and different amplitudes which are generated in the instrument (or externally). One signal frequency is low (from 60 to 400 Hz) and the other is high (from 1 to 12 kHz). The amplitude ratio of the two signals is about 4 to 1: the lower frequency being about 12 db stronger than the higher frequency. This composite signal is coupled to the amplifier input. If the amplifier is "clean," only these two signals will appear at the amplifier's output. If IM-D is present, the two signals will heterodyne and produce two additional frequencies—a sum and a difference.

To find out how much IM-D is added to the original input signal by the amplifier, the low-frequency signal is first eliminated by a filter in the analyzer. Then the total amplitude of the mid-range frequency is measured. This mid-range audio signal is then sent through an AM detector. The detector output provides the amount of IM-D, which is also measured. This measured IM-D is compared to the total amplitude of the mid-range frequency at the detector input and the ratio of the two amplitudes provide the IM-D percentage.

Of course, there are variations in the aforementioned design features. And the most usual frequencies used to check IM-D are 60 Hz (easily and economically available from an instrument power supply input) and 6 or 7 kHz. Under tighter specifications, the high test frequency used is equal to half the maximum specified frequency response of the amplifier and the low test frequency would be the lowest frequency response specified. For example, if an amplifier is said to have a response from 30 Hz to 20 kHz, the test frequencies would be 30 Hz and 10 kHz.

FLUTTER METER

"Wow," "flutter," and "drift" are triplets from the same dam. For practical purposes, then, we can group them under one name—flutter. This is especially desirable since neither the word "wow" or "drift" describes the basic effect as well as flutter does. And the effect is mechanical in origin. We will briefly touch on this subject again in those Chapters covering tape transports and phono turntables.

Many audio technicians questioned during recent times have said that they get excellent "by-ear" results when checking tape recorders, tape players, and turntables by employing pre-recorded test tapes and records having a low percentage of flutter. Some also have said that certain flutter cases could have been solved easier and faster with a flutter meter.

The flutter meter is an electronic test instrument designed primarily as an OEM (original equipment manufacturer) quality-control tool. It also is used as a recording and broadcast studio accessory. The instrument is designed to detect speed variations caused by mechanical defects in tape recorders, players, record players, audio-visual equipment—including so-called "sound movie" equipment. By measuring the flutter frequency, causes of flutter usually can be pinpointed to a specific mechanical fault.

Although we will not go into details here concerning the flutter meter's operation (refer to a particular instrument's technical manual), the basic flutter effect becomes essentially a form of FM at frequencies near 3 kHz. Hence, an FM detector or discriminator is used in the flutter meter to detect the modulation. A limiter-amplifier stage is used also to eliminate AM interference which would prevent an accurate measurement. The NAB (National Association of Broadcasters) standards call for flutter not to exceed 0.2% for recording and playback on the same recorder.

NOISE LEVEL METER

We spoke earlier about a noise- or audio-level meter. Whether called "audio" level, "sound" level, or "noise" level, this instrument can be used effectively for measuring overall background noise in various areas where PA and music-distribution systems are to be installed. Like all other test instruments, it should be used carefully, intelligently, and according to the manufacturer's instructions.

CHAPTER 3
AM Receiver Maintenance

Effective AM radio receiver servicing is a broad subject, and many of the basic troubleshooting procedures outlined here are fundamental to other electronic equipment. When you announce in your neighborhood that you are in the business of repairing home and commercial audio equipment, then you must be prepared to handle everything from table-top AM, AM/FM to other multi-band types, plus radios in TV-phono combinations, tape-recorders, tape players, home intercoms, phonographs, and auto radios. We will begin with basic trouble-shooting fundamentals and proceed to specific AM radio problems.

BASIC TROUBLESHOOTING TECHNIQUES

Keeping in mind the principles outlined in our modus operandi, let's look at a block diagram representing the basic circuits in both electron-tube and solid-state radios, as shown in Fig. 3-1. Notice that we have combined the oscillator and mixer stages into a single "converter" stage to simplify explanations. Separate oscillator and mixer arrangements will be discussed later. Notice also that we show an RF stage in the block diagram, but many small home radios do not have an RF amplifier stage. An AM radio generally has one or two IF stages. And the power supply may originate from an AC line source, a DC line source, or from a DC battery pack.

When an AM radio, of whatever type, comes into your shop for repair, 99 times out of a 100 it will be either (1) dead, (2) weak (or weak and distorted), (3) distorted with normal or almost normal volume, (4) noisy, (5) humming or motor-

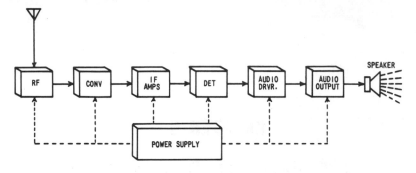

Fig. 3-1. Block diagram of a typical AM receiver.

boating, or (6) possessed with an intermittent condition characterized by one or more of the first five conditions mentioned above. None of these problems is difficult to solve—if you know the causes for the various symptoms—and they can be easily learned.

Let's take the dead radio first, and in so doing refute that often repeated "old-wives' tale" that dead radios are the easiest to repair. True, there's only one trouble symptom—a dead radio. But the number of possible causes and the time required to check the causes make a dead radio by far the potentially most difficult trouble-fault you can experience. We will concentrate specifically on a solid-state radio here, although similar troubleshooting procedures also can be applied to electron-tube radios. The schematic of a small solid-state radio is shown in Fig. 3-2. An electron-tube schematic is shown in Fig. 3-3.

DEAD RADIO

We have a dead radio on our hands, but our ears tell us that. With the cabinet removed to expose the radio chassis, we now "look," "smell," and "feel." We look for broken connections, especially open grounds, burned, discolored or charred resistors, plug-in transistors or other components not seated in sockets. We smell to detect burning resistors. We feel cautiously with a finger for overheating transistors or resistors. Although the "smell" and "feel" techniques are seldom used on small, low-current radios, these techniques are important on tube-type and more powerful solid-state radios.

These are the first fundamental steps in a proper troubleshooting procedure—we call it "sense inspection," or if you please, "sense perception," and the procedure applies to all electronic equipment.

We turn our attention first to the power supply and its associated components — including the on/off switch which is quickly checked with an ohmmeter—after the power supply is completely disconnected from the radio.

SIGNAL INJECTION

If the power supply and its components check all right and the on/off switch is not defective, we may inject a signal from a noise generator or other amplitude-modulated signal source at the wiper terminal of the volume control, while the control is adjusted full on. If nothing is heard in the speaker, the speaker is then quickly disconnected and a similar type is substituted by using clip-leads.

But suppose we do hear a strong signal in the speaker with a noise signal injected at the volume control wiper terminal, but no stations can be heard across the entire tuning dial? This tells us, of course, that our trouble is in a stage before the volume control, in the detector, the IF, the converter, or the RF stage.

WHERE NOW?

When we arrive at this point, and under these conditions, differences of opinion crop up among skilled technicians regarding the next troubleshooting move. One "expert" will inject an IF signal (455 kHz in home radios or 262 kHz in auto radios), modulated about 30% with a 400-Hz audio tone, at the detector input to check the detector. (NOTE: Some AM/FM/SW radios may have a different IF for AM/SW.) Then, if the detector is working, the technician proceeds with similar signal injections through the IF stages and to the converter output. At this point he would normally switch the signal generator to a higher frequency, probably to 1 MHz (1000 kHz) and then inject the signal into the radio's antenna through an induction loop. Thus, in this example, our technician has isolated the defective stage when the test signal disappears at the input of that stage.

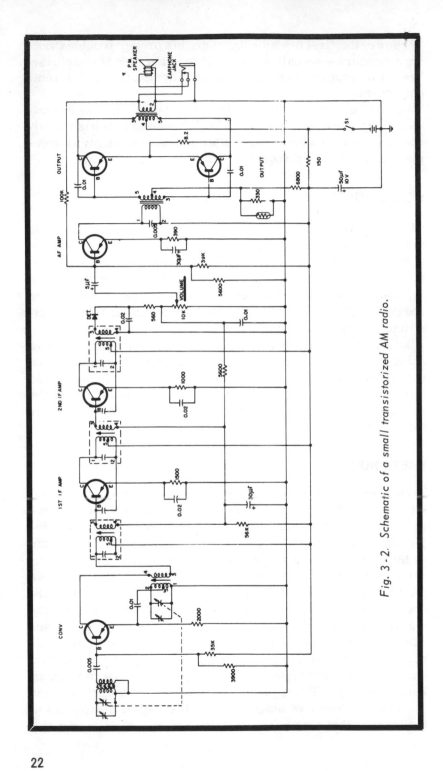

Fig. 3 - 2. Schematic of a small transistorized AM radio.

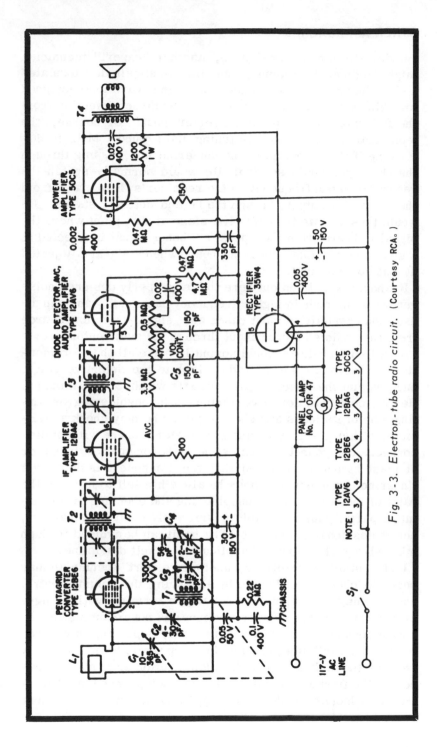

Fig. 3-3. Electron-tube radio circuit. (Courtesy RCA.)

OTHER APPROACHES

Under similar circumstances, another "expert" technician might begin at the antenna, injecting an amplitude-modulated 1-MHz signal into the receiver through a wire loop designed for this purpose, and then proceed by direct injection from the RF stage output to the converter input. Of course, the technicial would "rock" the tuning dial around 1 MHz to determine if the signal is being converted and passing through the IF to the audio section. He would normally switch the generator signal frequency to the receiver's IF to check from the converter output through the IF and detector stages. He would thus isolate the defective stage at that point where the signal is heard from the speaker when a signal is injected at a given stage output, but is not heard when a signal is injected at that stage's input.

It should be pointed out here that a properly designed noise generator is rich in higher frequency harmonics and can be used successfully in most cases as a signal-injection source from the radio's antenna input through the audio output.

A skilled technician would normally make all the aforementioned checks for a dead radio, isolate the defective stage, and pinpoint the defective part or parts within 5 to 10 minutes. He may confirm a defective component by unsoldering (or cutting) one of its leads and checking it (or he may check a transistor with an in-circuit transistor tester), but he will not disconnect a component lead unless he is thoroughly convinced (or at least strongly suspects) that the component is defective. Under no circumstances does he allow himself to fall into the habit of indiscriminately unsoldering and checking or substituting parts. This is a habit to which only confused amateurs, not professionals, succumb. The habit can be avoided through a knowledge of theory and application of self discipline.

Under similar conditions, another "expert" technician may begin to check a dead radio by advancing the volume control and touching the volume control wiper terminal with a finger or a screwdriver. (The screwdriver is not recommended on small-signal transistorized equipment.) Depending on whether or not a buzz or "click" is heard in the speaker, he immediately isolates the fault to either the audio section or to stages preceding the first audio stage. He may then proceed by various means to isolate the defective stage and defective component or components.

Let's go a little deeper into various checks which still another "expert" technician may make here to isolate the fault to a stage or component. Referring to the schematic shown in Fig. 3-1, if this technician hears an audible buzz or click from the speaker when checking with a finger, screwdriver, or an injected signal at the volume control wiper terminal, he immediately forgets about the audio section and turns his attention to the detector, IF, converter (or oscillator and mixer) and RF stage.

If he does not hear an audible click or buzz in the speaker, then he would normally check the on/off switch next. If the switch is functioning, he might then check the radio's total current drain by placing a milliammeter across the on/off switch with the switch open (solid-state radios only). He would use the milliammeter function of his VOM for this. If the radio's total current drain exceeds that specified by the manufacturer's service data, he would immediately check for shorted components (usually capacitors) connected from DC power leads to ground. If everything checks normal here, however, he would probably check the audio input and output transformer coils. Then he would substitute another speaker as described previously.

If the trouble is not located by now, our technician may check the emitter, base, and collector voltages on the audio output and audio driver transistors—using a manufacturer's schematic or service data to determine proper voltages. If the emitter and base voltages are low, but the collector voltage is normal, he would then check the supply voltages to the base and emitter at the input points of other capacitors and resistors. If the voltages are normal at the resistor and capacitor inputs but are still low directly at the base and emitter, he suspects a defective transistor. He would check transistors quickly with an in-circuit transistor tester before replacing them, however.

If this technician discovers immediately by the simple finger or screwdriver check that the audio section is good, he may temporarily ignore the second detector diode (which rarely fails except for poor solder connections), and go directly to the IF and RF sections, checking for signal continuity. If he finds these circuits are normal, he may then check the converter transistor's emitter, base, and collector voltages. If the base and collector voltages are low, for example, he may

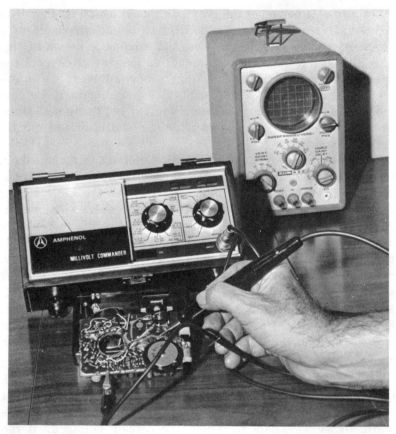

Fig. 3-4. When making bias voltage readings on transistors, use a meter having a scale capable of displaying as little as 0.2V accurately.

check the antenna coil winding for continuity. And if there is continuity, he suspects the converter transistor and would check it with a transistor tester. In many cases, the converter transistor will prove defective.

Suppose our technician finds the converter transistor's base and emitter voltages are close to normal but the collector voltage is low? Glancing at the schematic we can see that he would then check the oscillator coil secondary. If the voltage is normal at both ends, he would again assume that the converter transistor is defective. But if normal voltage exists only at one end of the oscillator coil secondary, he would correctly assume that the oscillator coil is defective. If the voltage is perceptibly low at both ends of the coil, he assumes that the IF transformer is defective since the DC voltage is routed through the IF transformer primary. When measuring transistor bias voltages, remember to use a meter capable of displaying 0.2v accurately (see Fig. 3-4).

As indicated at the beginning of this book, every person who wants to become an expert troubleshooting and repair technician will avoid trying to find "formula-type" do-it-yourselfer (DIYer) solutions to problems. No effective DIY-type troubleshooting "charts" have ever been designed except those which the individual professional technician has developed as a guide for his own use through a knowledge of theory and practical experience and he has this information filed away in his memory. Based on a knowledge of circuit theory and state-of-the-art troubleshooting practices as outlined here, the wise technician will develop his own particular procedures—a system which, at every step of the way, allows freedom of thought and experience. This is the "thinking" technician's approach to troubleshooting. And any "symptom-and-cause" charts which we may use in this book are intended as helpful guides. They can't be used as replacements or substitutes for "brain-gain." But we'd better get back to our dead radio before someone else fixes it.

If a radio passes no signal (no stations can be heard) but background noise comes from the speaker, then we can generally suspect that all sections of the radio are working, except either the antenna input circuits, the converter, or oscillator tuning circuits. In this case our skilled bench technician would check the ferrite-loop coils and the oscillator coil primary for continuity. He would also disconnect the oscillator tuning capacitor from ground and check it for shorts. These capacitors, because of very narrow spacing between rotor and stator plates, can short throughout their entire tuning range. Also, padder and trimmer capacitors can develop a short. We can now go still deeper into some problems which many technicians find especially difficult.

CONVERTER, OSCILLATOR / MIXER PROBLEMS

As we already know, the oscillator and mixer stages, either separate or combined into a "converter" circuit, perform the well-known superheterodyne function. The sine-wave oscillator generates a signal which is always maintained 455 or 262 kHz <u>above</u> (higher in frequency) the incoming broadcast station frequency as the oscillator tuning capacitor (ganged with the RF or antenna tuning capacitor) is varied. The oscillator signal and the incoming RF signal are fed to the mixer stage or to the mixer section of a pentode tube or transistor. Here, both signals beat together and produce two other signals—the sum and difference of the RF and oscillator signals. One of these two signals, the difference signal, remains constant as the tuning capacitors are varied throughout their ranges. We are not concerned here with the sum signal. The fixed difference signal contains the modulation information from the broadcast station. This is the signal which is fed to the stationary-tuned IF amplifier stages.

To repeat, what we call a "converter" is a single stage which combines both the oscillator and mixer functions either by using one multi-grid tube or one transistor. It is important that we understand the basic characteristics of these circuits if we are to successfully negotiate some particularly difficult problems which often arise in troubleshooting radios.

We are not concerned here with the design details of the three common oscillator types (Colpitts, Hartley, Meissner, and Armstrong) which may be used in radios. We are solely interested in fault symptoms, their causes, and how to isolate trouble in these circuits. And the two faults we're concerned with most are common to all three usual circuit types: A dead oscillator or an off-frequency oscillator. Either fault will prevent normal reception.

TROUBLESHOOTING CONVERTERS

You can approach these sections of an AM radio in various ways. When your dead radio checks out through signal injection from the speaker back to the first IF input, the next step is to determine if (1) the oscillator is operating and if so (2) is it on or off frequency? You can determine if the oscillator is operating by a number of different methods. Some have been recommended in the past which we feel have disadvantages. These include using a scope to check if sine-wave oscillations are being produced by the receiver's oscillator, also to measure the oscillator frequency by employing Lissajous figures.

It also has been recommended that another AM receiver be

used to determine if the oscillator of a dead radio is functioning. We consider this an amateurish approach, especially since the dials on most ordinary home-type radios are not accurately calibrated and the dead set's oscillator frequency cannot be checked accurately. The margin for error is also great—especially if the checks are attempted near a number of broadcast stations transmitting on different frequencies.

Another method frequently recommended involves using a VTVM to measure the voltage drop across the oscillator emitter resistor. The oscillator tuning capacitor stator and rotor plates are then shorted with a clip lead. It is said if a "change" takes place in the voltage being measured by the VTVM, then the oscillator is operating, the voltage change indicating that the oscillator stopped when the capacitor plates were shorted. The slightest loading effect on tube or transistor oscillators often prevents them from operating and meter indications are not always conclusive—especially in transistor oscillators. If you become sufficiently skilled, however, any one of these methods can be used effectively.

We would like to call attention to one other method which we believe is faster, more conclusive, and more accurate. It tells if a radio's oscillator is operating and on what frequency —doing both jobs simultaneously. And then if trimmer capacitor adjustment will not correct an off-frequency condition, you know immediately that a fault exists in the capacitive components of the circuit. This method requires a "grid-dip" oscillator. But when we mention a grid-dip oscillator, we do not mean the type generally used for "amateur" purposes. We mean a commercial-grade instrument which is both sensitive and accurately calibrated.

The job is done in two ways: While the radio is switched off, the "dip" function is used to see if the oscillator circuit frequency is correct by holding the instrument's coil-probe close to the radio's oscillator coil and tuning the instrument to obtain a minimum "dip" on its meter. The frequency is then read directly from the dial. In the same manner, while the radio is switched on, the oscillator action and its frequency can be determined by tuning the instrument for a maximum reading on the instrument's meter. Of course, a switch on the instrument is thrown to the proper function in each check.

"TOUGH DOG" TYPE PROBLEMS

Now suppose we find the oscillator is working and on frequency, but we still get no stations anywhere on the tuning dial? This situation seldom arises but it does happen. A number of unlikely possibilities can exist: (1) The oscillator's output amplitude can be very low. (2) Defective coupling from the oscillator output to the converter input (usually to the emitter of a converter transistor and to element 1 of a pentagrid tube converter). (3) The RF (or antenna) tuning capacitor is shorted or not tracking with the oscillator capacitor. (4) The ferrite-loop antenna coil is open at the converter transistor (normally the base or at tube element 7 in a pentagrid converter). (5) The tube or transistor is defective. How do we approach a problem like this?

SUBSTITUTE OSCILLATOR

We previously mentioned a few checks which can be applied to some of these situations. But let's try substituting another oscillator signal for the set's local oscillator. If this works, we can isolate the fault to a smaller area. Set the RF generator dial to a point 455 kHz higher than a local broadcast station which is known to be on the air. Say, a station operating on 610 kHz, for example. The RF signal generator's dial would be set at 1065 MHz. Now feed a signal from the RF generator output through a 0.01-mfd capacitor to the converter transistor's emitter or to element 1 of a pentagrid tube converter. Rock the dead radio's tuning dial back and forth around 610 kHz while adjusting the generator's output level up to maximum.

If the station is now heard in the speaker, you know that the radio's oscillator output is low or something is wrong with the coupling arrangement. At the same time you will learn that nothing is wrong with the RF signal input circuits up to the converter and it will be unlikely that anything is wrong except in the oscillator circuit coupling to the mixer section.

If this doesn't work, we'll have to check the voltages on the converter transistor and possibly the transistor's condition. Also check the tracking of the RF tuning capacitor. An accurate grid-dip meter comes in handy here again to compare the RF (or antenna input) signal circuits against the oscillator circuit frequency. As you already know, the RF circuits should be tracking 455 kHz lower than the oscillator circuit at all points.

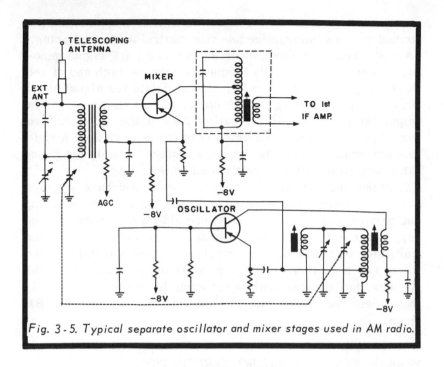

CHECKING OSCILLATOR / MIXER STAGES

Little difference exists between checking a combined oscillator and mixer section in a converter stage and checking separate oscillator and mixer stages. But it may be helpful to visualize the two states separately. An oscillator and mixer arrangement is shown in Fig. 3-5.

Now, suppose we have determined by any one or more of the various methods previously described that the oscillator is operating properly but no stations are heard throughout the tuning range of the radio dial (535 kHz to 1605 kHz). The skilled technician would probably go directly to the mixer transistor with a noise generator or 455-kHz RF signal modulated at 400 Hz about 30%. This signal is coupled through a 200-pfd capacitor alternately to the mixer transistor's base and collector. Because we expect a mixer stage to amplify a 455-kHz signal from 5 to about 15 times, the generator signal should be louder in the dead radio's speaker when injected at the base than when injected at the collector. If not, we would suspect the transistor. Remembering, however, that the average ear is not a very good indicator of amplitude changes within a narrow range, we do not suggest that the student or "raw" apprentice use this method in the beginning.

A more accurate check involves our trusty RF signal generator. Set the generator's frequency near the high end of the BC band (from 1 to 1.6 MHz) and modulate the signal about 30%. On radios having no RF stage, couple the generator's output through a small coil (about 3 or 4 turns of insulated wire 2 to 2 1/2 inches in diameter) to the dead radio's ferrite loop antenna. On sets having an RF stage, couple the generator loop to the mixer-input transformer.

Tune the receiver to whatever frequency you have selected at the high end of the BC band and rock the receiver's tuning dial around this frequency. If no signal is heard in the speaker, then the mixer is not operating properly. Check the transistor or the tube, the transistor or tube voltages, and the mixer transformer. If a good signal is heard from the speaker, but stations still do not come in on the radio's dial, then inject the signal into the antenna to check out the RF stage.

WEAK OR WEAK AND DISTORTED RECEPTION

Why do we specify "weak or weak and distorted?" For the simple reason that we can have a weak signal without distortion (insufficient gain for a certain set of causes) and a weak signal with distortion (insufficient gain for certain other different causes). As examples: (1) A weak signal (without distortion) can be caused by loss of gain in an RF stage (2) A weak signal (with distortion) can easily be caused by a leaking or open capacitor anywhere between the detector output and the audio output (see Fig. 3-6), or by a defective speaker. The amount of distortion will depend on a variety of factors, including the degree of leakage in a capacitor, combined with the particular design characteristics of a given circuit.

In most cases, weak but undistorted reception is caused by a sharp reduction in the RF signal supplied to the detector. Weak and distorted reception is usually caused by a fault in the audio circuit which attenuates the signal and upsets the circuit's linearity. As another example, a run-down battery in a transistorized radio can cause weak and distorted reception, primarily because of incorrect bias voltages on transistors. The same problem can develop in electron-tube radios because of incorrect bias or B-plus voltages and nonlinear circuit characteristics which arise because of off-value voltage-divider network resistors, leaking or off-value capacitors in coupling, bypass or decoupling networks.

Open electrolytic capacitors frequently cause weak and distorted reception, and the best way to handle these cases is by quick, direct substitution—using short clip-leads to "jump," or "bridge," the suspected capacitor. But one end of a leaking electrolytic should be disconnected before bridging. We also must remember to place the substitute capacitor in the same polarity as the original. An open, shorted, or leaking

Fig. 3-6. Weak and distorted signals are frequently caused by a defective electrolytic coupling capacitor connected from the volume control to the first audio stage.

AVC bypass capacitor can also cause distortion and weak signals.

Improper RF tracking can cause weak but undistorted reception; so can a broken ferrite antenna rod or a loose, improperly positioned coil on the rod. Although most modern ferrite antennas have been improved and will not present a problem of this type, you may run into one of the older types. For an accurate check on the proper position of the secondary coil on an older type ferrite-loop antenna, use two rods about 3 inches long—one ferrite and one aluminum rod. This is quickly done as follows:

Tune in a station near the middle of the BC band. Hold the ferrite rod close to the antenna, then take it away and hold the aluminum rod close to the antenna. In either case, the signal should attenuate sharply. If not, then the secondary coil should be positioned at a spot where each rod causes maximum signal attenuation. The coil should then be cemented at this point. Do not depend on your ears when making this adjustment. Use an AC VTVM across the speaker voice coil. A loose, improperly positioned coil can easily cause a marked reduction in signal strength as the meter will indicate.

Finally, a weak but undistorted signal can be caused by misalignment of either or both the RF and IF stages. And the IF stages are always aligned before the RF alignment is made. Step-by-step alignment procedures are furnished by all U.S. and most foreign manufacturers.

DISTORTION — NORMAL VOLUME

We will not discuss the general problems of audio distortion which arise under conditions of normal or near normal volume. But we will not be concerned with hi fi quality reproduction at this point. We are talking about audio distortion in small radios which can easily be detected by most human ears. To eliminate this kind of distortion we begin first by disconnecting one lead from the suspected speaker and substituting it with a known good one by using clip-leads. If there's a NC (normally-closed) earphone jack on the radio, you can plug the substitute speaker into it.

If this does not take care of the problem, "jumper" all electrolytic capacitors quickly. If no favorable results are obtained, check both halves of the input transformer secondary for balanced resistance (using the ohmmeter function of a VTVM) each side of center tap. Do the same for the output transformer primary. If a difference exists in the readings, substitute the transformers.

In transistor radios having push-pull outputs, insert a milliammeter (or microammeter on very small radios) in series with the base leads of both output transistors. Current readings here will vary from 200 to 500μa in most smaller radios, but we are not concerned with the amount of current. We are interested only in current balance between both bases. For example: The average small radio will show about 350 μa in either base leg of a push-pull amplifier running at maximum volume. If the current in one transistor base reads substantially higher or lower than this, suspect that transistor, but check it on a beta type in-circuit transistor tester before substituting. Current readings should be similar in both transistor bases.

In AC-operated electron-tube radios where one or more IF tube screens are bypassed through power supply output filters, a defective electrolytic can cause signal distortion. The usual tell-tale symptom here is a sharpened passband, causing the stations to tune sharply and distort.

NOISY RADIOS

It is not easy to define the various categories of noise without writing another book. But we are interested here only in those unusual reception-disrupting noises that originate inside the radio because of some component or circuit fault. For instance, a cracked resistor can generate noise. A failing capacitor can do the same. So can "cold" or "rosin" solder joints and power leakage between printed circuits caused by corrosion. If background noise suddenly becomes greater than it normally is in a given operating location, this can be caused by a loss of RF gain or misalignment of IF stages.

It takes a little experience to accurately diagnose the various causes for noise. But experienced bench technicians soon learn to tell the difference between one type of "frying" noise caused by a leaking IF transformer and that caused by a power leakage between printed circuits—after they have "tapped" and "poked" around printed boards for a few years with an insulated prod.

Many defective, "noise generating" capacitors, resistors, and poor solder joints can be detected by carefully tapping or gently pushing on them with a stiff, sharp-pointed plastic prod. But this must be done carefully to isolate the exact defective component or joint. Also, when re-soldering a suspected cold- or rosin-joint where a capacitor lead is connected, make sure the joint is at fault—not the capacitor. On many occasions a suspected sensitive joint is soldered and the noise disappears. But actually, the heat generated when the joint is soldered, temporarily "heals" the defective capacitor connected at that point. The capacitor will usually break down again shortly and resume its noise-generating antics. Defective tubes and transistors can also generate noise.

Isolating noise to a given stage or area is not an easy task at first but it becomes easier with experience. Many expert bench technicians find a sensitive scope useful for isolating noise to a given section—RF, converter, IF, or audio stage. The noise is thought of as an injected signal and is signal-traced and isolated to the stage of origin.

HUMMING AND MOTORBOATING

Hum is almost always confined to AC-operated radios and originates in the power supply. It is either 60 Hz or 120 Hz, the former coming from power supplies having half-wave rectifiers and the latter from power supplies having full-wave rectifiers. It is caused by defective filter capacitors. Cathode-heater leakage in AC-operated electron-tube radios can also cause 60-Hz hum.

Power supply hum is easily isolated by using a scope. If the ripple on the B-plus supply is higher than normal, then the power supply filters are "jumpered" with a known-to-be-good electrolytic, while observing results on a scope. If very little ripple is present, then start at the speaker output with the scope and trace back, stage by stage, until a point is reached where the hum disappears. Hum can be picked up by the signal circuits from an improperly dressed lead or through a shielded lead which is improperly grounded. Hum can also be caused by defective tube sockets.

Regeneration, spurious oscillations and "motorboating," is a trouble category which can be wide-ranging in origin. Changes in screen bypass capacitor values can cause regeneration. In those AC-operated tube ı .dio circuits where the screens are bypassed through output power supply filters, an open filter can cause motorboating—oscillations which are generally within a low frequency range. Here again, a substitute electrolytic is "jumpered" across the suspected capacitor by using short clip-leads.

It is generally understood that solid-state radios are more susceptible to motorboating, or oscillations that cause high-pitched squeals, than electron-tube radios are. Be this as it may, we can tabulate a list of causes for this symptom—rundown batteries, defective electrolytics, defective IF transistors, for example. Leakage or shorts between windings of IF transformers and open bypass capacitors can also cause squeals. High-impedance ground connections, like those frequently caused by rosin or cold joints, can set off oscillations. And finally, a defective transistor, particularly an IF transistor, can cause oscillations. In this case, generally, if the emitter of the transistor shows a higher-than-specified voltage, suspect the transistor and check it on a transistor tester.

INTERMITTENTS

Of all problems that arise in electronic equipment, the intermittent frequently is the most exasperating. Because it is a "here-today-and-gone-tomorrow" characteristic, the intermittent is often considered "difficult" by many technicians.

Generally, an effective instrumentation approach to intermittents is elaborate, complicated, and expensive. But some recent progress has been made in this direction. An intermittent is basically a "make/break," or "off/on" problem, and it can originate anywhere in radio circuitry from the antenna to the speaker—including any individual component in between. Finally, we have "cold" intermittents, "hot" intermittents, "vibration-sensitive" intermittents and, while rare, just plain "nonchalant" intermittents. But intermittents are not quite as difficult as some technicians have been led to believe. At least, on an average, no more difficult than problems which arise in the other five fault-symptom categories previously described. It depends again on how well we follow our modus operandi—plus a few special approaches.

To begin with, we have the intermittent that reveals the fault-symptom immediately when the equipment is switched on but disappears after the set heats to a certain point. This is a "cold" intermittent. But let's clearly understand what we mean here. We could use any one of the five previously described fault-symptoms as examples but let's use the first fault-symptom on our list—a dead radio. When the radio is first switched on it does not operate. But if left on it will begin operating when it warms to a certain point. And the time required to start operating may vary widely.

The "hot" intermittent does not show a fault-symptom when the equipment is first switched on. For example, take the second fault-symptom previously described—weak or weak and distorted reception. When the set is first switched on it operates normally, but if left on it will become weak or weak and distorted. And again, the time required for the reception to become weak or weak and distorted will vary.

The "vibration-sensitive" intermittent appears only when the radio is jarred, or vibrated. And the "nonchalant" intermittent is one that responds to nothing except perhaps shock from a sudden higher-than-normal voltage. And once again, although some of the five previously described fault-symptoms may be more prevalent among one or the other of these four intermittent types, all five can be intermittent.

ISOLATING INTERMITTENT FAULTS

To isolate intermittent faults, we must first study each intermittent type, learn to identify each type, and learn the various techniques involved in isolating the fault area and pinpointing the exact component or circuit fault. Let's first understand that the basic causes—the root fault—underlying all four intermittent types are similar. For example, although a "cold" intermittent is caused by an "open" or high-resistance connection that "tightens up" by expansion through heat, the root fault can include hair-line breaks in printed circuits, defective tubes, corroded sockets or tube base pins, poor lead connections inside capacitors, resistors, transistors, or other components. A cracked resistor can be another source of trouble. A "hot" intermittent is basically similar, except it operates in reverse—the contact opens or increases resistance because of or coincidental with an increase in temperature. Hence the root faults are the same. We will give examples shortly.

Our approach to intermittents is basic: We begin with a

thorough "sense inspection." And we use a good bench light, a magnifying glass, and a stiff but thin, sharply pointed insulated prod. If nothing shows up after sense-inspection, then we think once again briefly of the exact fault symptom (one or more of the five fault-symptoms previously described).

Assume the radio is dead when first switched on but after remaining on for a period it begins to operate normally. This frequently happens to electron-tube radios but very rarely to solid-state radios. And, if all the tubes light when the radio is first switched on, chances are the radio has a defective tube, especially in the power supply. If not, the problem could be caused by a leaking or otherwise defective electrolytic capacitor in the power supply or a defective voltage-divider resistor. It could be many other things also, from a cold solder joint in the converter tube circuit to a rosin joint on the audio output tube socket.

Let the set cool, switch it on again and quickly check B-plus voltages at the power supply output. If these are normal, inject a noise-generator signal at the volume control wiper arm. If you get a good signal from the speaker, move quickly to the last IF plate, to the grid, to the next IF plate and so on from stage to stage until a point is reached where the injected signal is no longer heard from the speaker. That's your defective stage, of course.

But how do we isolate an intermittent component? We'll see in a few minutes. But first, suppose the set begins to work normally before we isolate the defective stage? We switch the set off and let it cool. Then we begin with signal injection again where we left off when the set began to operate.

Once you've isolated the defective stage, and assuming the tube in that stage is known to be good, let the radio "cook" until it begins to operate. Now spray each component and connection in that stage one at a time with a coolant. When you spray the defective component or connection, the set will quickly go dead again.

This is also the general procedure to be followed with a "hot" intermittent, except that the set will come on and begin operating normally when you spray coolant on the defective part or connection. When using a coolant under these conditions, only a brief, quick spurt is necessary at the intended spot, otherwise you may cool one or more other nearby components or joints. Remember this is important: Make sure of the defective component or connection before you start removing parts or resoldering joints.

Although it can become extremely frustrating at times, we need say only a few words about the "vibration-sensitive" intermittent. Here, the radio goes on or off when moved or when tapped lightly. While it is operating, locate the defective component or connection by tapping and "poking" it with an insulated prod until the most sensitive point is located. It will always be, of course, a poor connection either in the circuit wiring or within a component, or corroded or insufficiently firm contacts in a battery container. Sometimes it's a cold or rosin solder joint. And not too infrequently the trouble can be caused by a defective capacitor, a coil, or a transistor. Perhaps the most troublesome fault is a poor connection inside a component—especially an IF transformer or transistor.

Various test instruments can often be used to aid in pinpointing a defective part or joint. For example, a voltmeter can be shunted across a resistor while its leads are prodded gently. A milliammeter can be placed in series with a transistor element and so on.

The "nonchalant" intermittent, while rare, can be approached only in a somewhat "violent" manner. For example, most capacitor testers provide an "overload" voltage function for checking intermittently defective capacitors. Because various capacitor types can break down and then "heal" temporarily, a good capacitor, rated at 500 working volts DC (WVDC), will usually take double that amount of DC voltage without breaking down. A defective capacitor of the same rating, however, will usually break down when the rated (or slightly more) DC voltage is applied. If you are unable to isolate an intermittent capacitor by previously described methods, try this capacitor tester "overload" approach.

Another unconventional method which sometimes takes care of a "nonchalant" electron-tube radio intermittent, involves a variable-voltage transformer (VVT) in the AC line input. The traditional VVT can be used effectively to reveal faults that show up on lower-than-normal line voltages. The process calls for slowly adjusting the AC input to the receiver from approximately 117v downward toward zero. Many ailing local oscillator circuits, primarily because of a defective tube, will frequently show up in the range from 90 to 117v.

40

It is not recommended, however, that a VVT be used to make intermittently defective components break down by boosting the potential beyond 117v, although this technique has been employed successfully by some technicians to isolate defective components.

Troubleshooting procedures can be speeded somewhat by using a few combination trouble-analyzing test instruments, including a form of intermittent "detector" designed to be used in conjunction with a scope. Although most of these instruments have been oriented toward TV service, some have been designed to aid the radio receiver servicing process.

KNOW — AND KNOW THAT YOU KNOW

A mistake that beginners make in many technical areas is to "assume to know." But this mistake is often made by those who have already had considerable, although insufficient, technical training. We'll give an example, and whether you work as an apprentice, an experienced technician or you're the boss, the example is valid in all cases. Many service-dealers have learned that some form of "on-the-spot" small radio repair can be profitably done by their particular operations. This system is especially helpful in battery sales. An important additional benefit is a healthier customer-relations image.

If you are sure of your knowledge and abilities, a special counter can be provided for doing this work right in front of the customer. This "in-the-open" method provides more benefits. But if you use this system, make certain that you know—and know that you know—the particular radio you are about to work on in front of the customer. If you are not thoroughly familiar with a particular radio, take it to the back service area, open it, and check it. You cannot afford to indicate to a customer, by the slightest hesitation or a baffled look, that you are not thoroughly familiar with his radio. He knows nothing about the various not-so-obvious methods employed to remove back covers, remove chassis, or replace batteries. And never use a schematic or service instruction booklet while working in front of a customer.

On-the-spot repair is concerned primarily with checking batteries, replacing tubes, repairing broken leads, and cleaning poor contacts in battery holders. You should also

Fig. 3-7. Speaker cones are easily punctured by a finger!

Fig. 3-8. Be careful with ferrite loops, they crack easily.

use your best customer-relations judgment regarding repair of very cheap transistorized radios. Remember, too, all batteries should be checked under load, with the on/off/volume control adjusted full up, or if the battery is removed from the radio, place a resistor across it (about 100 ohms for each volt), while making the voltage measurement. A 1K resistor is about right for a 9v battery. And if an old battery measures more than 20% below normal (about 7.2v for a 9v battery), the customer should be advised that "the battery is getting low."

But it should be understood here that a battery in this condition can give reasonably good service on some radios for a considerable additional time if the radio is operated intermittently for brief periods, giving the battery a chance to "rest" for long periods. It should also be pointed out that when a battery's voltage is measured while it is in the radio, the tuning dial should be set off-station. The voltage of even a new battery can easily drop 20% or more on some radios when tuned to a strong station at full volume.

If not properly instructed, one of the first mistakes that beginners make is to push a finger-nail or the entire end of a finger through a speaker cone (see Fig. 3-7). Be careful with speaker cones and ferrite-loop antennas (see Fig. 3-8). When removing a small radio chassis from a case, be careful not to lift the chassis by holding on to components. Always handle the chassis by the outer edge of the printed board. The work bench area for small, plastic-cased radio repairs should be covered with a heavy felt cloth which has been securely anchored to the work bench top.

When a customer complains about the battery "not lasting very long," use the 50 ma function on your VOM and measure the total current drain of the radio. With the radio off, place the meter probes across the on/off switch. Depending on the radio's power, current consumption under normal conditions may range from 10 to 25 ma or more—under no signal conditions. If the current drain exceeds the manufacturers' specifications, do not replace the battery until the component causing the abnormal current drain has been replaced.

CHAPTER 4
Working With FM Tuners

As emphasized throughout this book, we are concerned here only with basic principles—the important things involved in troubleshooting and repairing FM tuners. Once again, we will avoid rehashing and repeating technical information which concerns only design and research engineers, information which is frequently found in scores of technical books ostensibly prepared to "train" and inform electronics technicians.

FM TUNER PRINCIPLES

It is assumed that you are already familiar with the principles of amplitude modulation (AM) in radio broadcasting and reception. AM and FM can be compared, in a simplified way, by visualizing AM as "vertical stretch" and FM as "horizontal stretch." If you have doubts about the practical validity of these definitions and comparisons, glance at the AM and FM modulation waveforms shown in Fig. 4-1A and B, respectively. As you can see, the AM carrier's amplitude varies up and down vertically when modulated by a sine wave, and the FM carrier's frequency varies from side to side horizontally when modulated by a sine wave. Although the "frequency stretch" of AM and FM cannot be directly compared, except in an oversimplified way, both systems are limited by Federal Communications Commission (FCC) regulations.

By this late date, most every one must now be familiar with the unusual advantages offered by FM. The very nature of the present FM spectrum (88 to 108 MHz), plus the limiting characteristics of the reception method, reduces or totally eliminates much ordinary radiomagnetic interference (including that from local electrical storms). Tuning is sharper, too. And little or no "skip-distance" interference is experienced from distant broadcast stations as in the AM spectrum, especially at the higher frequency end of the AM band. Briefly, FM is capable of providing higher "FI" than AM, hence one important reason for the present growing popularity of FM.

As in AM, the superheterodyne principle is employed in FM tuners. But here, essentially, the similarity ends. The FM IF is 10.7 MHz as compared to the 455 and 262 kHz found in most home and auto radios, respectively. When we look at an FM tuner, we immediately notice that the coils are smaller and have fewer turns of larger-gauge wire, the tuning capacitors have fewer and usually smaller plates, the fixed capacitor values are smaller and connecting circuit leads are shorter and more rigid.

The physical differences between FM tuner components and those in AM tuners immediately indicate the higher-frequency nature of FM equipment. It also becomes obvious that the nature of FM equipment demands higher level, more exacting troubleshooting and servicing techniques.

FM DETECTORS

Because the intelligence in FM broadcasts is contained in

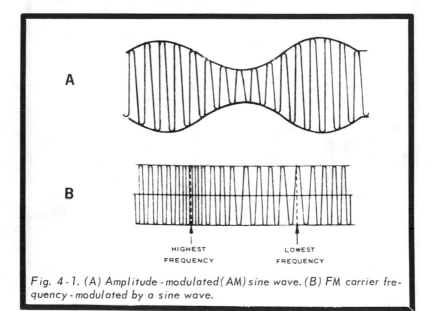

Fig. 4-1. (A) Amplitude-modulated (AM) sine wave. (B) FM carrier frequency-modulated by a sine wave.

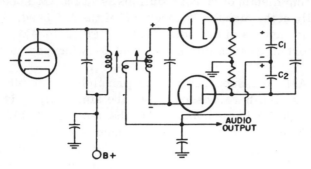

Fig. 4-2. Schematic of an electron-tube ratio detector.

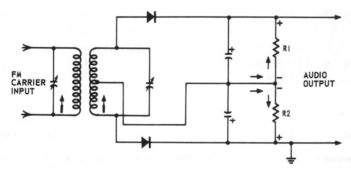

Fig. 4-3. Solid-state discriminator-type FM detector circuit.

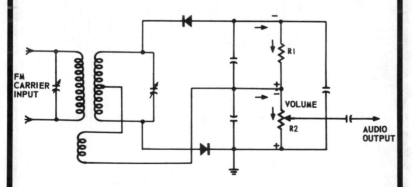

Fig. 4-4. Solid-state ratio detector FM detector circuit.

the carrier <u>frequency</u> variations instead of carrier <u>amplitude</u> variations as in AM, our FM detector is designed to convert these frequency variations into intelligible audio at the receiver output. Although a wide variety of different circuit arrangements can be employed in FM detectors, including the well-known Foster-Seeley "discriminator," the combination detector/limiter 'BN6 and 'DT6 electron-tube circuits and other arrangements, we will be concerned here primarily with the conventional ratio detector, an electron-tube version of which is shown in Fig. 4-2.

DISCRIMINATORS vs RATIO DETECTOR

First, let's attempt to clear up some confusion which occasionally arises regarding FM discriminators and FM ratio detectors. Some technical writers have incorrectly used the words interchangeably. The discriminator type FM detector, not widely used today, functions like an "error detector" or phase detector used in some TV horizontal control circuits and in many industrial feedback-loop control systems. Notice how the diodes are connected in the FM discriminator circuit shown in Fig. 4-3 as compared to the diodes in the ratio detector shown in Fig. 4-4. Notice also that the voltages across R1 and R2 in the discriminator oppose each other while in the ratio detector the resistors function in series. The audio output from the discriminator is across both resistors while in the ratio detector it is across one resistor.

It is important to remember that both detector-types have no output when the transmitted carrier is at rest, or at its exact center frequency. But the most important thing to remember is: All types of FM detectors produce the same basic results—a varying audio signal at the output—and the major service problems these circuits present to practical technicians are also similar. It should be understood also that the FM discriminator-type detector is extremely sensitive to AM signals (essentially noise impulses), and requires one or more limiting (or "noise clipping") stages to reduce AM signals. The ratio detector, properly designed, requires little or no limiting.

As in all inexpensively designed consumer products, most lower priced FM tuners (especially those in combination AM/FM portable and table-model home types), incorporate many cost-saving circuit innovations. These include combined detector/limiter arrangements, and in some designs the final IF stage provides a certain amount of limiting. Other design economies also are employed. And some modern FM tuners, designed essentially for stereo reception, have various other noise-limiting systems which use transistors and special silicon diodes. For example, one high-quality FM tuner designed for an FM/stereo multiplex demodulator employs one diode and two transistors in a Schmitt-trigger noise-limiting circuit.

AUTOMATIC FREQUENCY CONTROL

Before we go into troubleshooting, repair and alignment problems peculiar to FM tuners, it is important to know that the RF local oscillator frequency must be stabilized. If it drifts even a small amount, the IF will shift sufficiently to cause severe distortion or a complete loss of reception.

Although there are a number of ways to stabilize the oscillator frequency, the most common arrangement uses a voltage from the ratio detector or discriminator output which is fed to a reactance tube or a "varactor" diode circuit. These circuits are, in effect, shunted across the oscillator tank circuit and act as automatic variable capacitances. It may prove helpful in troubleshooting to have a cursory understanding of the simple basic operating principles involved in these AFC circuit arrangements. It is assumed here that the reader has a thorough knowledge of AC fundamentals.

The simplified circuit shown in Fig. 4-5 is a triode electron tube reactance-type AFC circuit. The oscillator tank circuit, L1 and C1, is shown in the dotted box. The values of R1 and C2 are selected so the current across R1 and C2 leads the oscillator's output voltage by 90°.

With the reactance circuit drawing a 90° leading current from the oscillator output, the tube becomes a capacitive load across the oscillator output—the capacitance increasing with a current increase and vice versa. FM detectors are normally designed to provide the AFC circuit with a positive voltage when an increase takes place in oscillator frequency and a negative voltage when the oscillator frequency decreases. As previously mentioned, a positive voltage will increase the tube's plate current and in turn its shunt capacitance, which reduces the oscillator frequency. Of course, the reverse effect takes place with a negative input to the AFC circuit. All components in an AFC circuit are critical, especially R1, C2, cathode bias resistor R3, and cathode bypass C4. The amount of voltage on the tube plate is also important.

VARICAP AFC CIRCUITS

As previously mentioned, many FM receivers today employ silicon varactor (variable-capacitance) diodes in AFC circuits. Here, the diode is reverse-biased—the higher the reverse bias, the lower the capacitance. Additionally, some circuits use a zener diode to stabilize the bias on the varactor. And, as in all AFC circuits in this area—whether tube or solid-state—an RC network must be used between

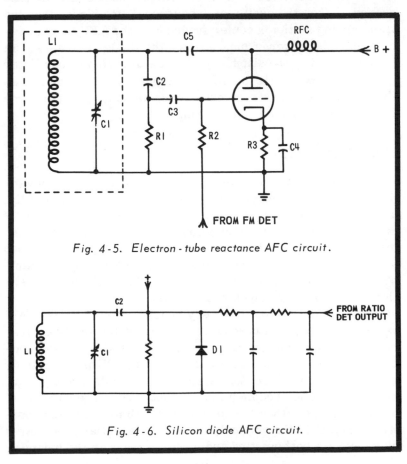

Fig. 4-5. Electron-tube reactance AFC circuit.

Fig. 4-6. Silicon diode AFC circuit.

the ratio detector or discriminator output and the AFC circuit input to prevent the circuit from responding to the regular audio variations in the FM detector.

A simplified AFC varactor circuit is shown in Fig. 4-6. The effective components include the varactor, D1, capacitor C1 shunted across the oscillator coil, L1, and series capacitor C2. The reverse-biased diode acts as a capacitance shunted across the coil. In a simplified way, the circuit works as follows: The diode cannot conduct because it is reverse biased. But its bias varies with the varying voltage from the discriminator or ratio detector. If the oscillator frequency drifts lower than the transmitting station's center frequency, the AFC voltage increases, the reverse bias voltage across the diode decreases, in turn reducing the shunt capacitance across the oscillator coil which raises the oscillator frequency. When the oscillator frequency drifts higher than the transmitting station's center frequency, the AFC voltage decreases, the reverse bias across the diode increases, in turn increasing the shunt capacitance across the oscillator coil which lowers the oscillator frequency. Knowing something about how these circuits operate may provide some helpful clues when troubleshooting.

LIMITER CIRCUITS

As previously stated, a properly designed FM ratio detector does not require conventional limiting stages. Generally, adequate limiting can be provided by designing the last IF stage to operate as a combination amplifier/limiter. But the discriminator type detector requires limiter stages. And since you will occasionally come across FM tuners having discriminator-type limiting stages, we will briefly discuss conventional-type limiter stages at this point.

Because FM detectors are very sensitive to all types of AM variations, including unequal frequency response of receiver circuits, atmospherics and "man-made static," it was necessary to design noise limiters, "clippers," or "amplitude gates." Even ratio detectors require some degree of limiting if top-grade reception is to be obtained. The purpose of a limiter stage, then, is to equalize both negative and positive "stretch" in the vertical direction. Essentially, a limiter stage cuts off both positive and negative peaks in the incoming signal at a certain predetermined level. This is illustrated in Fig. 4-7A and B.

Plate and screen voltages of some electron-tube limiters are relatively low—with the screen voltage usually about half that of the plate. But we will not go into the detailed theoretical design-oriented operation of these circuits. The principles involved in conventional electron-tube and transistor limiters are similar—limiting action being obtained by designing and biasing the circuits so both positive and negative amplitude variations are maintained at a constant level by alternately driving the tube or transistor into cut-off and saturation.

A conventional single stage pentode electron-tube limiter is shown in Fig. 4-8 and a transistor limiter stage is shown in Fig. 4-9.

TROUBLESHOOTING FM TUNERS

Once again, we visualize the FM tuner as a whole unit, made up of a few, simple individual circuits as seen in the block diagram shown in Fig. 4-10. We must learn what the

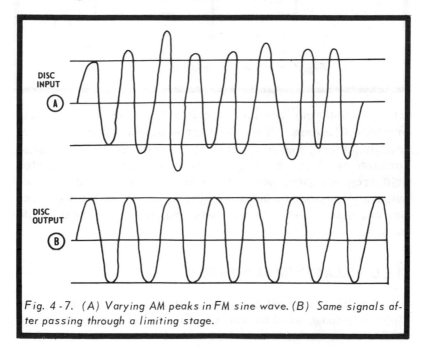

Fig. 4-7. (A) Varying AM peaks in FM sine wave. (B) Same signals after passing through a limiting stage.

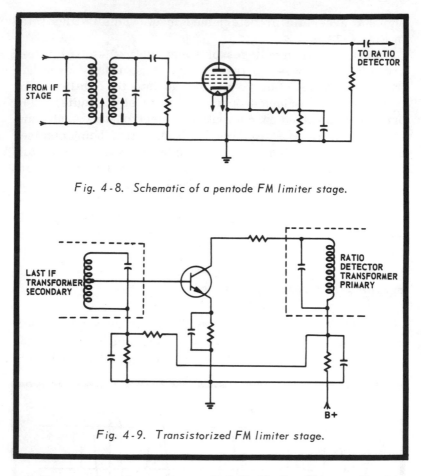

Fig. 4-8. Schematic of a pentode FM limiter stage.

Fig. 4-9. Transistorized FM limiter stage.

trouble symptoms are for each individual unit as outlined in previous Chapters. This is learned primarily by technical training, study, and practical experience. Some of the most common problems arising in FM tuners include oscillator drift, regeneration, poor AM limiting, or noisy and distorted output. A number of other symptoms arise but we will confine our discussion here to these and some of their possible causes.

THE OSCILLATOR OR CONVERTER

Basically, FM oscillator and mixer or converter stages are similar to those used in AM receivers. The only difference is that these circuits are designed to operate at higher frequencies. The oscillator difference frequency (IF) is higher, too—10.7 MHz instead of 262 or 455 kHz as in standard AM auto and home receivers. And the number-one fault-symptom in FM tuners is probably <u>drift</u>. But it should be understood that most FM oscillators will drift some—even in expensive consumer equipment—up to 20 or 30 minutes after the equipment has been switched on. This sometimes causes distortion, necessitating retuning the receiver to a given station's center frequency, depending on the strength of the signal being received. Incidentally, this fact should always be explained to your customers.

Why don't we digress here momentarily and point out one theoretical fact which, if thoroughly understood, will frequently make servicing easier? All sine-wave oscillators, no matter what type, are Class C <u>amplifiers</u> having resonant (regenerative, or positive) feedback. To put it another way, "an oscillator is an amplifier is an oscillator." When we get to the problems of regeneration and its causes in FM receivers, you will be wise to remember this technical fact. Additionally, by disconnecting the feedback loop in an oscillator, we can check the circuit by using the same techniques employed in checking amplifiers. Hence, in servicing FM oscillators we think of them as having two parts: the amplifier section and the in-phase feedback section.

Because of the frequencies involved, most FM oscillator drift is caused by temperature changes surrounding the oscillator's frequency-determining components. And we must not

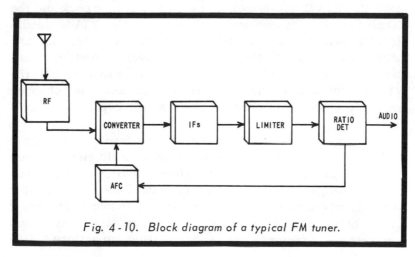

Fig. 4-10. Block diagram of a typical FM tuner.

forget that a defective AFC circuit also can cause oscillator drift, either to a higher or a lower frequency, or it can be intermittent in either direction. Some FM oscillator circuits employ small - value temperature - compensating capacitors across the frequency-determining elements, or tank circuit. These can develop defects and cause drift.

CHECKING FOR DRIFT CAUSES

Determining the cause or causes of oscillator drift can frequently become a long-winded, tedious job, especially if the drift is intermittent. Just as frequently, however, the "pain" can be eased if we use our heads—knowledge and experience. For example, suppose we have a drifting oscillator which we believe might be caused by a defective, ordinary GP (general purpose) capacitor having a tolerance of plus or minus 10%.

We can approach the problem in many ways. But we'll discuss only two general approaches here. Perhaps the easiest way is to check all capacitors in the oscillator tank circuit by using a good capacitor tester having provisions for making an "over-voltage" check for breakdown.

Another way is to connect a zero - center - reading VTVM (previously adjusted to zero-center) across the receiver's detector output. Switch the receiver on and tune a moderately strong FM signal to center frequency by observing whatever tuning indicator the receiver affords. If the VTVM needle swings either right or left, adjust the oscillator tank slug until the VTVM reads zero. Let the receiver cook on the bench for at least one hour. Observe if the VTVM needle drifts gradually in the negative or the positive direction and notice how much. Now switch the receiver off and let it cool for one hour. Repeat the aforementioned process. If the VTVM needle drifts in the same direction in each case, chances are your suspicions were correct: a bad capacitor. Let's find out.

Let the receiver cool again for an hour (and with the same VTVM set-up) then hold a hot soldering iron tip close to each capacitor for a few seconds in turn (not counting temperature-compensating capacitors) and observe the VTVM needle. If it suddenly moves in the same direction as previously described in the heat-run checks, there's your defective capacitor. But what about temperature - compensating capacitors? Use your head: If the soldering - iron check on the GP capacitors shows no sudden change in the VTVM readings, then the chances are you can forget those capacitors and look for a defective temperature - compensating capacitor. Warm up the capacitor tester or quickly disconnect one end of the TC capacitor and "tack-in" another of the exact same type and give the tuner another heat-run.

Isolating drift faults to the oscillator or AFC circuits is relatively easy for well-trained troubleshooters. Depending on the amount of drift, similar checks can be made with the AFC being alternately connected to and disconnected from the oscillator circuit while observing and comparing the magnitude of drift on the VTVM in each case. And if you employ your basic training and a little imagination, you can come up with a variety of innovations and new, more rapid approaches to the problem

REGENERATION

It appears that regeneration is more prevalant in FM receivers than in AM receivers. The reasons appear obvious when we understand the nature of VHF. Although the direct causes of regeneration symptoms are few, primarily because most symptoms are caused by both low- and high-order harmonics of the IF being radiated or "looped" back to the front end from limiter or detector stages, the symptoms are wide ranging. So are the paths that the signals causing regeneration can take.

Some regeneration symptoms include the well-known "motor-boating" effect. This may be a low-frequency oscillation or it may soar to a high-pitched squeal, and it may occur only at certain spots on the tuning dial. Regeneration can cause distortion which may be continuous, intermittent, or may appear on some stations and not on others. It can cause noise and "hiss." It can cause distortion on weak signals, "dead-spots" at various points of the tuning dial and other weird effects.

As mentioned previously, wiring and lead-dress is critical at VHF frequencies. When working on FM tuners, removing or replacing defective parts, it is extremely important not to disturb the manufacturer's original lead dress. Replacements should be wired in exactly as the original was wired, at the exact same position and having the same lead lengths. This is particularly true of lead dress surrounding a limiter or a detector stage in hand-wired tuners. This problem is less serious, however, in printed circuit (PC) wiring. But bypass capacitors in AFC, heater, and B-plus leads can fail in PC wiring boards and allow IF harmonics, sometimes going well beyond the 10th-order harmonic of 10.7 MHz, to feed back to the RF, oscillator, mixer, or converter stage, and cause the symptoms previously described. This may seem fantastic, but considering the frequency-distorting nature of FM limiters and detectors, it is easily understandable. As in all positive feedback from higher-level to lower-level stages, it does not take much signal in lower-level stages to cause disruption at the receiver's output because, as you already know, positive feedback reinforces.

SOLVING REGENERATION PROBLEMS

If you are thoroughly familiar with the nature of VHF and UHF, you already know that special precautions must be taken while working with this equipment. For example, if you want to check small ceramic bypass capacitors in heater, AFC or B-plus lines for "opens," you can't hold a substitute capacitor in your hand and expect to come up with accurate results. You tape the test capacitor (not with friction tape but plastic-type tape) to a plastic rod about eight inches long. The leads should be clipped short and should protrude forward so the suspected capacitor can be bridged by the one mounted on the rod. And make sure the bridging capacitor is the exact same type as the one in the circuit.

We should not forget, too, there also is one other source of regenerative feedback in FM tuners—from a defective capacitor in the power supply—especially those sets which use combination filter and IF screen-grid decoupling arrangements. This, however, has become much less a problem in modern solid-state equipment. A scope, substitute power supply, or a bridging job on power supply filters can quickly determine if the motorboating is caused by a defective filter located in the power supply or by a defective component located elsewhere.

In the event a tube-type tuner is being checked for regeneration, it should be cooked for an hour or two, checked out on both strong and weak signals and its AC power source should be boosted to approximately 128v with a variable-voltage transformer to make certain regeneration does not take place at higher input voltage levels.

INADEQUATE FM LIMITING — NOISY OUTPUT

Although regeneration can cause a noisy FM tuner output, the most likely cause is inadequate AM limiting. But, you may inquire, what causes inadequate AM limiting? The answer is: Any below-par or defective component from the tuner's antenna input to the detector output. And this includes misalignment in any stage from the RF section through the detector stage. A weak detector diode, especially in a discriminator-type detector, can cause poor AM limiting on weak signals. Noise will occur also if a ratio-detector circuit electrolytic opens. With some variations, normal troubleshooting procedures, as previously described in Chapter 3, under "Weak Signals In AM Receivers" is in order here. But one other highly desirable check should be made, especially after making extensive repairs on a tuner. Check the tuner's sensitivity. The subject is sufficiently important to deserve a modest amount of general treatment here.

TUNER SENSITIVITY

We are not concerned here with sensitivity measurements based on that signal strength required at the tuner input in microvolts (μv) to give a specified audio power output. This method of arriving at a tuner's sensitivity may work roughly for tuners that operate at much lower frequencies, but we want to know how little signal at the tuner input will over-ride the internal tuner noise sufficiently to produce a "clean," acceptable output which provides enough power to drive the preamp or amplifier which the owner is using. This is a particularly important consideration with equipment used in FM/stereo multiplex reception.

A variety of expressions have been and are still being used by manufacturers to indicate a receiver's sensitivity. This has caused considerable confusion in the minds of many technicians. Sensitivity has been expressed, for example, as a signal-to-noise ratio between the internal signal and

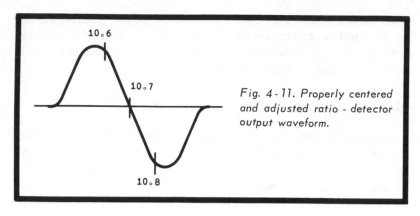

Fig. 4-11. Properly centered and adjusted ratio - detector output waveform.

noise level, like 10:1—apparently meaning 10 μv of signal to 1 μv of noise.

In other cases, a tuner's sensitivity has been rated as "40 db"—supposedly meaning the signal is 40 db above the receiver's internal noise. In still other instances, the sensitivity has been indicated by the expression "-40 db," again apparently meaning the noise is 40 db below the signal level or the signal level is 40 db above the noise. Nowadays, a manufacturer may say that his tuner will operate satisfactorily with a "minimum usable input of 2 μv," for example. And some manufacturers may say their tuners will provide a certain number of "db quieting with a certain μv input." A typical specification within a certain equipment price range may indicate "20 db of quieting with 0.7 μv signal input."

Whatever the method used to indicate a receiver's sensitivity, we want to check the equipment to see if it meets the stated specification. This holds true whether we are checking a separate tuner or a combination tuner/preamp/amplifier. We must be able to accurately inform the owner whether or not his equipment is operating within its design capabilities, in the event he has a complaint. He may actually require a better antenna installation. And, because of the wide variety of methods employed to express the sensitivity rating of FM receiving equipment, the particular checking method recommended by the manufacturer should always be employed. Insist on a block diagram of the test instrument setup, technical details, and step-by-step procedures.

Although a tuner's sensitivity is not based on the antenna installation, an inadequate FM antenna installation can cause you to suspect the tuner's sensitivity. And before you attempt a sensitivity check, make sure all components in the tuner are functioning normally and the tuner is properly aligned throughout, from the RF stage to the detector output. This subject is further explored in Chapter 5.

DISTORTED OUTPUT

We will not discuss here the type of distortion which arises in audio amplifiers. (Various phases of audio amplifier distortion are covered in Chapters 1 and 3, and the subject is further discussed in forthcoming Chapters.) Neglecting a particular type of distortion caused by too much ripple in the tuner's power supply, distortion or poor tone quality from an FM receiver can be caused by one or more faults existing in the tuner's local oscillator, a limiter stage, IF stage, or ratio-detector circuits. This includes defective limiter or IF tubes or transistors, incorrect tube or transistor voltages, misalignment of limiter or IF stages, open load resistors across IF transformers, open IF bypass capacitors, and unequal conduction in detector diodes—especially in discriminator-type detectors. Distortion from this source usually occurs on strong signals.

If a limiter stage does not receive enough signal to saturate, a form of distortion can result which would normally be caused by an inefficiently operating RF or IF stage. Of course, some previously mentioned faults, in conjunction with regeneration, can cause distortion, especially if an IF stage defect allows that stage to approach the oscillating point. This type distortion appears on weak signals but disappears on strong signals.

If a ratio detector or discriminator transformer is misadjusted (especially the secondary), severe distortion can result. If the primary is not properly adjusted, the signal will be weaker than normal and some distortion also may result. The cure for this type of distortion is detector transformer alignment according to the manufacturer's specifications. This is best done by using a sweep generator, crystal-controlled 10.7-MHz marker oscillator, an accurate AM signal generator and a scope.

The general procedure calls for injecting the sweep signal through a capacitor into the base of the last IF (or limiter) stage. If your sweep generator does not have an internal

10.7-MHz marker, clip the leads of a separate marker generator onto the chassis five or six inches each side of the ratio detector transformer. Adjust the top core of the transformer to provide an "S" curve having the 10.7-MHz crossover point exactly on the base line as shown in Fig. 4-11. Then adjust the bottom core so the top and bottom portions of the curve are the same shape and amplitude. Use your signal generator as a variable marker to check the linear limits of the curve which should not be less than 75 kHz each side of the 10.7-MHz center point, preferably 100 MHz as shown.

To obtain a full waveform on the scope, the sweep generator should be set normally to sweep a bandwidth extending approximately 500 kHz above and below the 10.7-MHz IF center frequency. This is a much better method than adjusting the bottom core for maximum signal output on a VTVM and then adjusting the top core for zero-center.

CHAPTER 5

Servicing FM Stereo Equipment

In addition to a ratio detector, an FM/stereo multiplex (mpx) tuner also has an mpx detector (demodulator) and other decoding circuitry. The theory involved with stereo/mpx, boiled down to simple terms, will be briefly outlined here. First, however, we need to clarify the meanings of two words: monophonic and stereophonic. The words monaural and binaural are sometimes incorrectly used to mean monophonic and stereophonic. The literal meaning of monaural is "one-eared," and the literal meaning of binaural is "two-eared." Mono and stereo, as we shall use them here, mean a single-channel audio system and a two-channel audio system, respectively—in both transmission and reception.

FM MULTIPLEX

A compatible FM stereo signal, as broadcast from an FM radio station, is essentially a mono signal when received on a mono-type FM receiver. But the composite signal is made up of left- and right-channel information from two microphones, a stereo phonograph pickup, or stereo tape head. The information from both microphones or other dual-source transducer is combined (L plus R) and fed to the FM transmitter.

Additionally, a 38-kHz subcarrier is amplitude modulated by a resultant signal derived by electrically subtracting one channel from the other (L minus R). This subcarrier is suppressed at the transmitter by a balanced modulator, and the unsuppressed sidebands also frequency-modulate the transmitter's carrier. The 38-kHz subcarrier is generated and reinserted at the multiplex receiver.

A third signal, a 19-kHz pilot (one-half the 38-kHz subcarrier frequency) also frequency-modulates the FM carrier a small percentage of the total modulation—not more than 10%. This pilot signal, like the 3.58-MHz color-burst signal transmitted by a color TV station, is used to sync the receiver's 38-kHz local oscillator. The 19-kHz signal is usually amplified and doubled to 38-kHz and reinserted in the multiplex receiver demodulator.

Finally, some FM stereo stations who sell background music and other services under a Subsidiary Communications Authority (SCA, or sometimes called subscriber carrier assignment) employ a second subcarrier at 67-kHz for "storecasting." This information is trapped out in the typical FM stereo receiver to prevent interference to FM stereo reception. It should be noted also that FM radio stations not equipped to transmit FM stereo, may also transmit SCA information—employing a subcarrier in the L minus R spectrum (23 to 53 kHz) for this purpose. FM stereo receivers are designed to either automatically or manually compensate for this by disabling the 38-kHz signal when receiving mono information from a station of this type.

When a regular FM receiver is used to receive an FM stereo broadcast, the information received is essentially mono, or L plus R, as previously stated. When an FM multiplex receiver is employed—containing not only a regular broad-band FM detector but also a mpx detector and other "decoding circuits"—the information received includes the L plus R signals, the L minus R signals, the 19-kHz pilot signal and the 67-kHz SCA signals. But the 67-kHz signal is trapped out before it arrives at the mpx detector. The L plus R and L minus R signals are combined in the mpx demodulator which provides a left- and right-channel output for stereo reception.

This explanation may appear oversimplified to some, but it is about all we need to know about FM multiplex to successfully troubleshoot and service the equipment. Other important theoretical details and clarifications which may appear useful are covered later. Remember, an FM stereo transmitter is modulated by at least three signals, sometimes four. Observe the transmitter block diagram shown in Fig. 5-1. Notice the two microphone inputs which are capable of picking up audio frequencies from 30 Hz to 15 kHz. Notice that both microphone outputs go to an "adder." The adder output is the L plus R signal which frequency modulates the carrier—this is the signal which is received by a mono FM receiver.

Observe also that the right microphone output goes to a phase inverter. This signal's phase is shifted 180° and is fed into another adder, along with the left-channel output. This adder's output is L minus R and is fed to the input of a balanced modulator. A 38-kHz oscillator voltage is also fed to the balanced modulator. This 38-kHz subcarrier is suppressed in the balanced modulator leaving two sidebands which carry L minus R information within a frequency range from 23 to 53 kHz (the sum and difference of 15 kHz, the highest audio frequency which the FM transmitter will normally handle). The demodulator output (L minus R) also frequency-modulates the transmitter's carrier. And finally, the 19-kHz "pilot" signal also modulates the transmitter's carrier. (We have ignored the SCA subcarrier generating equipment since it does not concern us here.)

What is known as the frequency spectrum of the composite

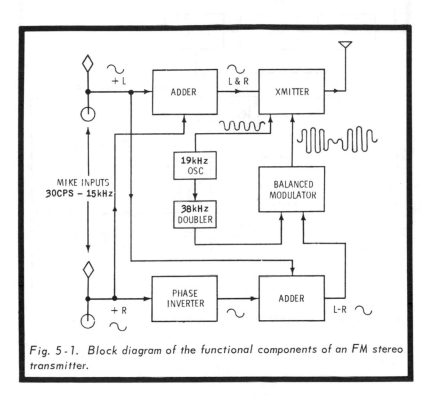

Fig. 5-1. Block diagram of the functional components of an FM stereo transmitter.

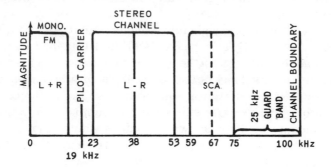

Fig. 5-2. Frequency spectrum of the composite FM stereo signal plus SCA.

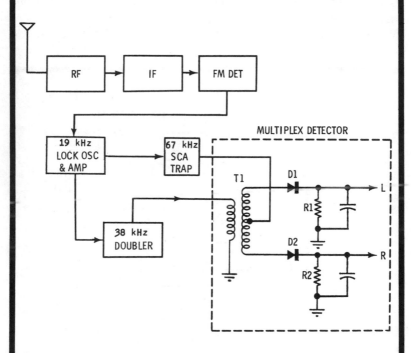

Fig. 5-3. Block diagram of an FM stereo tuner. (Courtesy Motorola)

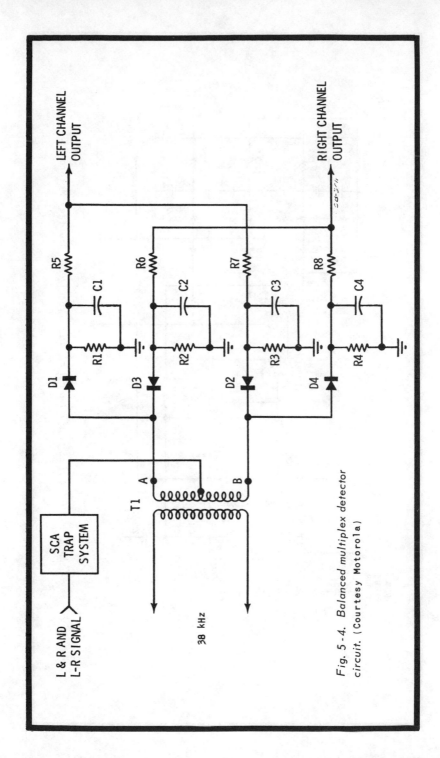

Fig. 5-4. Balanced multiplex detector circuit. (Courtesy Motorola)

65

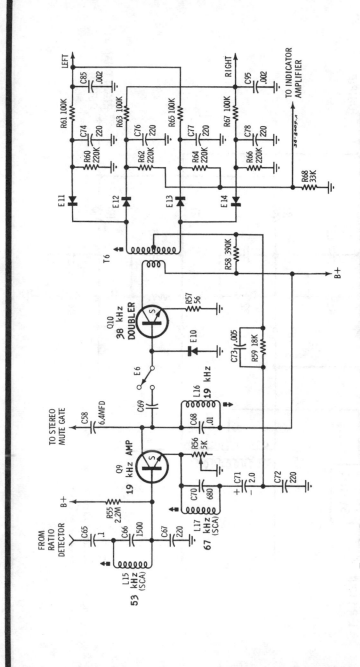

Fig. 5-5. Major multiplex circuits of a typical FM stereo receiver. (Courtesy Motorola)

FM stereo signal is shown in Fig. 5-2. It should be remembered, however, that the L minus R sideband spectrum from 23 to 53 kHz contains audio frequencies from 30 Hz to 15 kHz —the same as the L plus R signals.

FM MULTIPLEX RECEIVERS

The block diagram of an FM stereo tuner is shown in Fig. 5-3 which includes a simplified mpx detector (demodulator) schematic. A balanced synchronous detector is shown in Fig. 5-4. Although this detector is used in an FM receiver especially designed for mobile (specifically automobile) use, it is a good "typical" study-example of an economy-type modern solid-state FM stereo receiver. A schematic of the major mpx circuits are shown in Fig. 5-5.

Referring to Fig. 5-5, pay particular attention to the tuned SCA traps L15-C66 and L17 - C70 which are designed to effectively suppress SCA signals from 53 to 67 kHz. Also observe L16 and C68 which resonates the collector circuit of Q9 to precisely 19 kHz, the pilot carrier. The signal is coupled through C69 to Q10's base. Diode E10 is important since it and the base/emitter junction of Q10 conduct on alternate half-cycles—regulating and maintaining the 19-kHz signal at a relatively constant amplitude. T6 is tuned to 38 kHz, the second harmonic of 19 kHz, and it couples the signal into the multiplex demodulator.

Stereo information from the emitter of Q9 goes to the center-tap of T6 where it is compared to the reinserted 38-kHz subcarrier, and the left and right stereo signals are reproduced across load resistors R60, R62, R64, and R66. The 220-pfd capacitors bypassing the four load resistors filter out residual 38-kHz signals which are almost completely cancelled out by this type demodulator.

The two RC networks consisting of R61, R65, and C85 and R63, R67, and C95 serve to de-emphasize what the FM transmitter pre-emphasized in its audio section—a process which attenuates the lower frequencies more than the higher frequencies. These two networks, typical of similar arrangements in all FM receivers, are designed to "reverse" (de-emphasize) the transmitter process. Admittedly, this is a somewhat oversimplified explanation, since the process also eliminates considerable "noise" in the HF audio spectrum, but

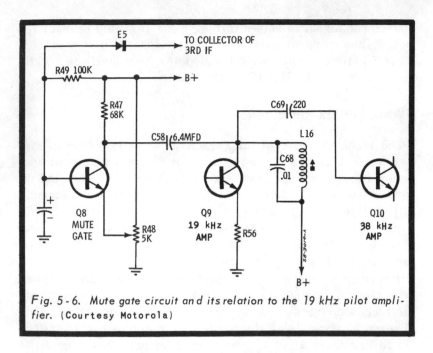

Fig. 5-6. Mute gate circuit and its relation to the 19 kHz pilot amplifier. (Courtesy Motorola)

from the troubleshooting viewpoint this particular subject isn't worth further discussion.

MUTE GATE

The "mute gate" circuit shown in Fig. 5-6 is, in effect, an automatic switch. When the incoming signal's amplitude is too low to provide good stereo reception, the radio switches to monophonic reception. It may prove helpful to understand something about this circuit's operation.

The mute gate is "tied" to the negative half of the 10.7-MHz IF signal from the third IF transistor collector. Resistor R49 provides forward bias to Q8. Diode E5 conducts through R49 on the negative half of the 10.7-MHz signal and E5's conduction varies with the 10.7-MHz signal amplitude. The more negative voltage developed by E5, the lower the forward bias provided through R49 and vice versa. This action will switch Q8 on and off. Since the collector/emitter impedance of Q8 is low when conducting and high when not conducting, it effectively controls the 19-kHz amplifier's action. When Q8 is cut off (strong input signal) Q9 and Q10 operate in a normal manner for stereo reception. When Q8 conducts (weak input signal), it disables Q9 and Q10 to prevent Q10 from amplifying noise.

As previously mentioned, the composite stereo signal normally appears in the emitter circuit of Q9 and when a mono station is being received, the 38-kHz doubler has no output. But under this condition, diodes E12 and E13 will be forward biased through resistor R58. These diodes will then pass the mono (L plus R) audio signal to the right-and left-audio sections.

Refer to the stereo indicator circuit shown in Fig. 5-7. The indicator lamp, E17, will glow when a stereo signal is received. A positive voltage is developed at the base of Q12 when diodes E12 and E13 conduct, providing forward bias to Q11 which amplifies the signal and switches Q12 on. The pot (R56) in Q9's emitter circuit is set to control the level at which the indicator circuit will "fire" when a stereo signal is received.

BALANCED MULTIPLEX DETECTOR

Although a two-diode multiplex detector is practical, a number of good reasons exist for using balanced-type (bridge) detectors which employ four diodes. Refer to the schematic shown in Fig. 5-4. Equal but 180° out-of-phase 38-kHz voltages appear at points A and B. When point A is positive, point B will be negative and diodes D1 and D2 will con-

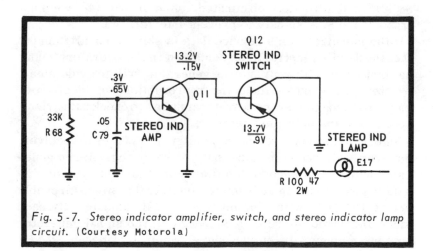

Fig. 5-7. Stereo indicator amplifier, switch, and stereo indicator lamp circuit. (Courtesy Motorola)

duct simultaneously. On the next half cycle, D3 and D4 conduct simultaneously. In both cases, equal currents always flow through both halves of T1's secondary. Remember: L plus R and L minus R information is fed from Q9's emitter circuit to T1's secondary center-tap. The 38-kHz signal is fed to the primary of T1 and is subsequently cancelled out through the balanced detector's action. In effect, this mpx detector operates like a high-speed switch—it "flips" and "flops" 38 kHz per second and "samples" a portion of the information contained in both right and left channels.

Two diodes select right-channel information and pause while the other two diodes select left-channel information. The audio section of this FM receiver appears in Fig. 6-30A and B in Chapter 6. The power supply is shown in Fig. 8-16, Chapter 8.

MULTIPLEX ALIGNMENT

The manufacturers' service data spells out the alignment method to fit his particular receiver design, even to the point of calling out the exact test points to attach the various test instruments. All we need to remember is: (1) The mpx detector must receive both L plus R and L minus R signals. (2) We must have a minimum amount of interaction (crossmodulation) between the two channels (left and right). (3) The reinserted 38-kHz subcarrier phase must match that of the transmitted 19-kHz pilot signal which is normally doubled to 38 kHz and ultimately eliminated. And finally (4) we must trap out the SCA signals to prevent interference.

If the manufacturer's service data is skimpy on alignment, then the detailed instructions contained in the operators manual that came with your stereo generator will provide step-by-step procedures—including reproductions of ideal-type, easy-to-recognize scope waveforms used to check out various sections of the multiplex decoder.

Before you attempt to check or align an FM stereo receiver, however, it is recommended that you study your stereo generator manual and become familiar with it and the type of signals it generates. In addition to separate RF/sweep (probably set at 100 MHz with plus or minus 5 MHz adjustability each side) and audio signals, a standard stereo signal generator puts out a composite stereo signal which includes a 38-kHz

L minus R signal superimposed on an L plus R signal, plus a 19-kHz pilot signal. The composite waveform from the generator as it appears on a scope or at the FM ratio detector output is shown in Fig. 5-8.

Depending on a particular manufacturer's instructions, we may begin by adjusting the SCA traps first. We assume (1) that the tuner is perfectly aligned and functioning normally from the RF input to the ratio or discriminator detector output and (2) that your stereo generator provides all the facilities and functions mentioned here. Use the representative block diagram shown in Fig. 5-9 as a rough guide.

Remove the speakers from both channels and substitute with proper load resistors (both value and wattage). Switch the stereo generator on and let it warm up for 15 or 20 minutes. Switch the receiver on and tune it to 100 MHz (or to a clear channel between 95 and 105 MHz), the stereo generator's RF

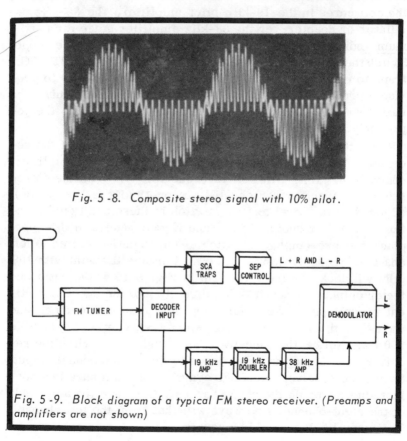

Fig. 5-8. Composite stereo signal with 10% pilot.

Fig. 5-9. Block diagram of a typical FM stereo receiver. (Preamps and amplifiers are not shown)

frequency. Remove the receiving antenna and switch the stereo generator to 67 kHz. Couple its RF output to the antenna terminals through a 300-ohm termination. Switch the receiver off and connect a scope or AC voltmeter at the output of the trap (or traps). Switch the receiver on and adjust the trap core for minimum deflection or amplitude on an AC VTVM or scope. Notice that some mpx decoders have a trap at 53 kHz which may or may not be adjustable. And other receivers may have an additional adjustable trap at 72 kHz. In these cases, inject signals at the proper frequency and adjust the trap cores as before to obtain a minimum indication. Switch the receiver off.

Connect an AC meter or scope to the output of the 38-kHz amplifier where it feeds into the demodulator and switch the stereo generator to 19 kHz (pilot signal frequency), set on the stereo mode. Switch the receiver on. Adjust, in turn, the coil cores in the 19-kHz pilot amplifier, the 38-kHz oscillator or doubler, or the 38-kHz amplifier stage for maximum indication and a clean 38-kHz sine wave on the scope. Switch the receiver off. At this point we have adjusted the SCA traps to eliminate interference and made adjustments to provide a proper 38-kHz signal to the demodulator input. The next step is to find out if the demodulator is doing the job properly.

Place the scope leads across the load resistor at the receiver's left-channel output. Set all preamp (or amplifier) controls to "flat." Switch the receiver on. With the stereo generator function switch still set on 19 kHz (stereo mode), adjust the pilot level to 10%. Switch in these additional functions: a 1-kHz audio signal, and if your generator has one, place the pre-emphasis switch to on, function switch to left channel, mode L minus R. (The L minus R signal with 10% pilot will look like the waveform in Fig. 5-10 at the ratio detector output or when it is fed directly into the scope's vertical sweep input.) We should now have a pair of perfect sine waves on the scope when its sweep frequency is adjusted to 500 Hz. Notice the sine-wave amplitude and switch the receiver off. Move the scope leads to the right-channel output. Switch the receiver on. Once again, we should have two perfect sine waves on the scope screen. Compare the amplitude of the right-channel sine wave with that from the left channel. Their amplitudes should be equal.

Move the stereo generator mode switch to L plus R (mono) and repeat the above steps for both channels. If the receiver has a separation control, adjust it so the sine wave at the left-channel output has the same amplitude as the one at the right-channel output.

Now suppose the sine waves at the receiver output are not good? What if they are misshaped, fuzzy, distorted, or clipped? Maybe they got that way going through the preamps or amplifiers. In this case, move the scope back to the demodulator output and repeat the checks to determine if the fault is in the decoding section or the preamp and amplifier sections. If the sine-wave output is good on one channel and not good on the other, this should give you some idea on how to proceed further.

CHANNEL - SEPARATION CHECKS

Because it is also an important consideration, perhaps we should mention the basic principles involved in making a channel-separation check. The Federal Communications Commission originally set certain standards for the compatible FM stereo system. One standard specified that the transmitting equipment must be designed and adjusted to provide at least a "29.7 db separation" between channels. Multiplex receivers are designed to provide 30 db or more separation. Hence, the receiver must be adjusted so no crosstalk exists between channels. In other words, the audio information in one channel should not "spill-over" into the other. If the 38-kHz os-

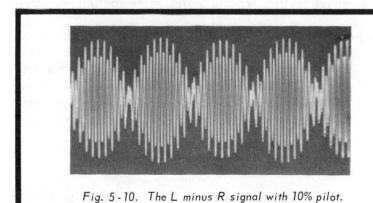

Fig. 5-10. The L minus R signal with 10% pilot.

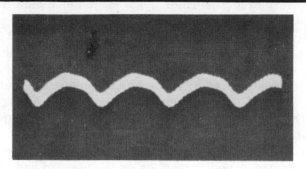

Fig. 5 - 11. This waveform indicates a no - signal condition at either right or left channel output with the signal to the opposite channel — an indication of considerable cross - talk between channels.

cillator (or doubler) is tuned to resonance and is on frequency and locked by the 19-kHz pilot signal, we should have no problem here—assuming the tuner is normal and the preamps and amplifiers are normal.

With the generator set to provide a stereo composite and a 1-kHz audio signal to the 100-MHz RF, the pilot level at 10% and channel switch left, connect the scope across the left-channel output load resistor. Observe if you have a good clean 1-kHz sine wave on the scope. If so, move the scope probes to the right-channel output. Under normal circumstances you would now see a waveform resembling that in Fig. 5-11, but thinner and having less p-p sine-wave amplitude. Perfect (ideal) separation would be indicated by a straight, thin line. At any rate, carefully refine the 19- and 38-kHz coil or transformer adjustments to obtain a straighter and thinner waveform.

Now place the scope across the right-channel output and switch the generator to right channel. Observe if you have a good 1-kHz sine wave here. If so, move the scope to the left-channel output. Again the scope waveform should look somewhat like the one on the left channel after the last adjustments. Carefully refine the 19- and 38-kHz coil slugs again to make the waveform still straighter and thinner.

db SEPARATION

Maybe you would prefer to go at channel separation in another way. With the stereo generator still set up as before, substitute your audio VTVM for the scope. Connect the meter to the right-channel output across the load resistor. Adjust all tone controls to "flat." Adjust gain and balance controls to provide a small reference voltage reading on the meter, say 1v full-scale (0 db on some meters). Notice the reading. Move the meter to the left-channel output and generator channel switch to left. If necessary, slightly readjust gain and balance controls to match the reference voltage established on the right channel. Recheck the voltage on the right channel after moving the generator channel switch to the right. If you have trouble balancing these voltages, then you may really have trouble. Otherwise proceed.

With the AC VTVM still connected to the right-channel output, switch the generator channel selector to the left and read the meter. This reading should be the channel separation in db. You probably will have to switch the meter to a lower scale, say 0.1v full-scale (-20 db). If the separation is better than that (and it should be), you'll have to switch to the 0.03v scale (-30 db).

We have given only a few of the checks necessary on FM stereo receivers, and this only in a very general way. But the procedures are not as complicated as they sound. In fact, with the exception of SCA trap adjustments, many skilled technicians use an "on-the-air" stereo signal and meter to make most alignments and adjustments. But we do not recommend this to anyone except the most experienced.

It should be clearly understood that we have been using a stereo generator with a built-in RF generator capable of being modulated by a variety of lower-frequency oscillators. Thus, the stereo generator output can be connected to the receiver's antenna input. This arrangement makes it possible, among other things, to use the composite stereo signal waveform to check out the overall condition of the FM receiver, including IF and ratio detector passbands, the preamp, and the amplifier, prior to making stereo decoder checks and alignments. Under normal circumstances, when checking the tuner with a composite stereo signal, the scope is placed across the ratio detector or discriminator signal output.

As another example of the experienced technician at work, he capitalizes on the fact that the mpx balanced demodulator has two separate halves and when trouble develops in one half, he frequently uses the functioning half by coupling a signal from it to the defective half through a 0.1-mfd capacitor connected to one end of a clip lead. Both halves of the balanced demodulator seldom become defective at the same time.

FM RECEIVER SENSITIVITY

As promised in Chapter 4, let's further explore the somewhat confused subject of FM receiver sensitivity. This is not an easy job, for a variety of reasons—primarily because industry standards vary. Some manufacturers who have taken their equipment ratings and specifications more or less seriously in the past, have ambiguously defined FM sensitivity "as the amount of signal required across the receiver antenna terminals to provide a stated signal level at the discriminator or ratio detector output." The test procedure goes like this: You inject a certain small amount of FM signal (usually modulated at 30% at the tuner's antenna terminals). Let's say, for example, to obtain an arbitrary standard output of 2 mw. The sensitivity rating is based on the number of input microvolts required to obtain this output. Whatever else this may be, it has the familiar ring of an ancient formula for determining receiver gain. The question has been asked repeatedly (without a satisfactory answer), what about the noise? Noise becomes part of the FM stereo receiver's "sensitivity" figure and the 2 mw previously mentioned could easily be 20% or more noise. In practice, of course, much larger signals are usually employed. Some manufacturers use very small signals. The average FM tuner may, for example, have an output anywhere from 200 to 500 mw.

In recent times some design engineers have apparently come to the conclusion that an "FM-quieting" rating is more significant. The approach here is slightly different. True, the same 30% modulated signal is employed to establish a reference level which includes both noise and signal. And this reference is obtained at a point below the receiver's limiting capacity, otherwise no "significant" figure can be obtained. Once again, the measurement is made at the detector's output. Then, ostensibly with a flourish, the 30% modulation is switched off and another measurement is made. We now come up with what appears to be a "db-quieting" figure which seems to be the ratio of the signal and noise level to the noise level. In high-class stereo equipment this db-quieting is normally specified from about 35 to 40 db or better. But wait! Not all receiver manufacturers do it this way. And many skilled technicians have been asking questions about this.

For example, how does a 30% FM test signal stack up against a 100% modulated composite FM stereo signal? What happens to the "db-quieting" when we use the composite stereo instead of the 30% sine-wave-modulated signal at the antenna input and make our measurements once again at the ratio detector output? The difference is not brain-shocking—but there is a difference. Every experienced technician knows by now that the s/n ratio is poorer on FM stereo than it is on FM mono.

Considering the overall situation being what it is, we must once again suggest that you ask the manufacturer how he arrives at tuner sensitivity and then use his method to check the tuner to determine if it comes up to specifications. Only then can you be in a position to intelligently advise the owner— your customer—if he complains about poor reception. He

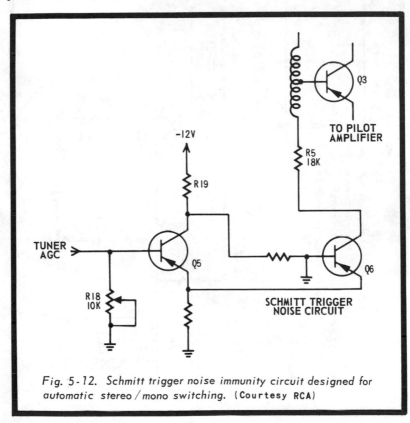

Fig. 5-12. *Schmitt trigger noise immunity circuit designed for automatic stereo/mono switching.* (Courtesy RCA)

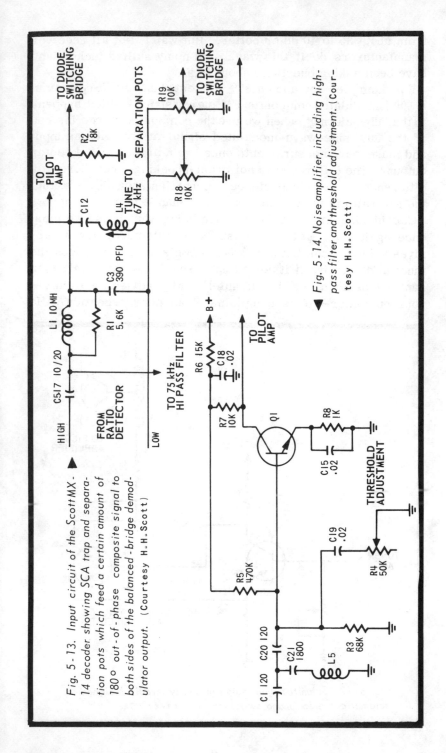

Fig. 5-13. Input circuit of the Scott MX-14 decoder showing SCA trap and separation pots which feed a certain amount of 180° out-of-phase composite signal to both sides of the balanced-bridge demodulator output. (Courtesy H. H. Scott)

Fig. 5-14. Noise amplifier, including high-pass filter and threshold adjustment. (Courtesy H. H. Scott)

may, and frequently does, have an inadequate antenna installation.

DECODER CIRCUIT VARIATIONS

Some years ago RCA designed a multiplex decoder which used a Schmitt trigger noise circuit for automatic stereo/mono switching. The 2-transistor circuit is shown in Fig. 5-12. The output of the trigger circuit is connected to the base of the 19-kHz pilot amplifier. The input is connected to a negative AGC source supplied by the tuner's third IF collector.

When a weak signal is being received, no AGC voltage is available to drive Q5 and it is switched off. Q6 is on. With Q6 conducting, Q3 is switched off because it does not receive enough bias voltage. This disables the 19-kHz pilot amplifier, no 38-kHz signal arrives at the balanced demodulator, and only L plus R signal is fed to the left and right channels.

When a strong composite stereo signal is received, AGC bias is developed and feeds the base of Q5, biasing it on. Q6 is biased off and the voltage rises, switching Q3 on through R5, an 18K resistor. Stereo operation resumes and a voltage from the 38-kHz amplifier switches on another transistor which lights the stereo lamp. R18 can be varied to set the "on" trigger level. The "off" trigger level is determined by the value of R19, which is approximately 8.2K.

Another interesting circuit is the one used by H. H. Scott in their MX-14 decoder. To begin with, the FM detector is not grounded, making two floating output connections available. These go to the SCA filter and balancing network. Here, the incoming composite signal is balanced to ground and a phase-shifting network which contains two pots is used to shift the composite signal 180°. A small portion of this out-of-phase composite signal, together with the regular composite signal, is fed to the left- and right-channel outputs of the demodulator diode switching bridge. This innovation provides a more complete cancellation of the unswitched portion of the signal, hence an aid to channel separation. These pots must be carefully adjusted for best results. A schematic of the SCA filter and balancing network is shown in Fig. 5-13.

Additionally, a noise amplifier circuit employs the incoming noise above 75 kHz for stereo/mono switching. The noise amplifier stage is shown in Fig. 5-14. Noise is fed through a

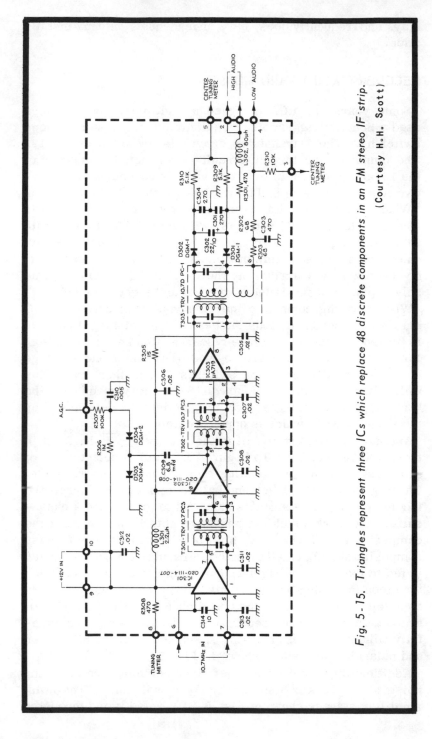

Fig. 5-15. Triangles represent three ICs which replace 48 discrete components in an FM stereo IF strip.

(Courtesy H. H. Scott)

a high-pass filter from the high side of the FM tuner output. This noise is amplified and then fed to a full-wave diode noise detector circuit. The DC component developed here is used as cut-off bias on the first pilot amplifier. R4 is a pot which can be adjusted to provide the amount of noise required to cut off the pilot amplifier. To properly adjust this control, the receiver is first tuned to an unused clear channel on the dial. The control is then varied until the demodulator switches to stereo. The adjustment is then backed off until it is just past the point where the demodulator switches back to mono. The other sections of the MX-17 are more or less conventional.

Additional circuits which we will be confronted with in today's FM stereo receivers are: between-station noise-muting circuits which provide silent tuning between stations, various types of tuning indicators, FETs (field-effect transistors) in preamps and tuner RF stages and integrated circuits in IF stages.

The schematic of an FM IF board containing three small IC packages together with the ratio detector is shown in Fig. 5-15. This board is used in the Scott solid-state AM/FM stereo receiver, Model 341. The IC unit, μa719, contains the equivalent of about 34 discrete components and is provided with eight connections. The other two ICs (301 and 302) also have eight connections and contain the equivalent of seven discrete components each.

CHAPTER 6

Preamps and Amplifiers

Although many electron-tube type amplifiers are still in use, much audio amplifier equipment made today is solid-state. Hence, this Chapter deals primarily with solid-state circuits. As in electron-tube amplifiers, it is sometimes helpful to be concerned with amplifier classes: Class A, AB, and B especially.

It is important to remember that we work with two general amplifier types—"small-signal" and "large-signal." The small-signal amplifier, of course, precedes the large-signal amplifier, most often in cascaded stages to build sufficient signal to properly drive the final, or output amplifier, which in turn drives the audio system's speaker or speakers.

You are already familiar with the three basic circuits used in solid-state amplifiers: the grounded (common) base, the grounded (common) emitter, and the grounded (common) collector. We'll concentrate on grounded-emitter and grounded-collector type amplifiers, since the grounded-base amplifier is presently little used in home- and commercial-type audio communications equipment.

GROUNDED-EMITTER AMPLIFIERS

The grounded-emitter amplifier, in whatever circuit configuration, is perhaps the most widely used transistor amplifier. A simplified schematic, employing an NPN transistor, is shown in Fig. 6-1. If a PNP transistor is used, the DC supply polarity would be reversed. Notice that the emitter is "common" to both the input and output. A voltage divider bias-stabilizing network (R1/R2) is used, in addition to a stabilizing resistor (R4) in the emitter. This is the only basic transistor amplifier circuit which reverses the signal phase 180° from input to output. How this phase "reversal" takes place is not important to us.

Let's assume that a circuit of this type or one having slight variations, is used as a low-level, small-signal, Class A driver stage in an audio amplifier. We recall from our studies of basic amplifiers that a Class A amplifier is biased to operate on the most linear portion of the dynamic transfer characteristic curve, as shown in Fig. 6-2. Resistors R1 and R2 are selected so their effective values (consistent with the input signal level), cause collector current to flow continuously during the complete AC input cycle. And some current flows even when there's no input signal.

Since R1 and R2 determine the bias voltage between base and emitter (disregarding temperature effects), their values are critical to the amplifier's proper operation. Resistor R4 is also critical and affects bias conditions. We must remember that the linear range of a transistor's operating curve is more

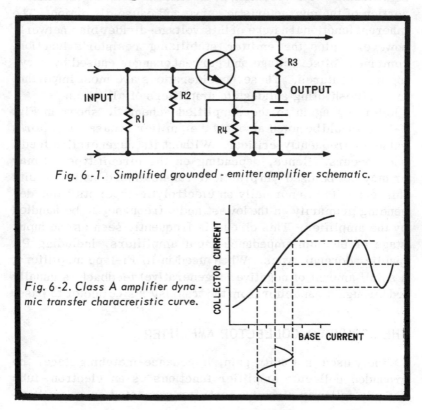

Fig. 6-1. Simplified grounded-emitter amplifier schematic.

Fig. 6-2. Class A amplifier dynamic transfer characteristic curve.

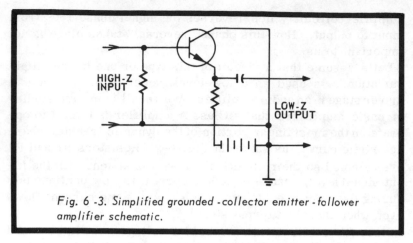

Fig. 6 -3. Simplified grounded -collector emitter -follower amplifier schematic.

narrow than on electron tubes. Additionally, the transistor's operating characteristics are more sensitive to changes, including temperature variations. If the resistor values change significantly, the operating point will move into a less linear portion of the characteristic curve and cause distortion. The inherent functional nature of this voltage-divider bias network, however, plus the emitter stabilizing resistor's function, combine to offset voltage and current changes caused by varying temperatures. These considerations are most important in troubleshooting solid-state amplifiers of all types.

Referring again to the simplified schematic shown in Fig. 6-1, it should be noted also that the emitter bypass capacitor's value is frequently critical. Without it, degenerative feedback occurs. Hence, depending on the circuit type, it may or may not be used. In practical transistorized audio circuits this capacitor is normally an electrolytic-type; its value depending primarily on the lowest audio frequency to be handled by the amplifier. This circuit is frequently seen as the input stage of medium-impedance input amplifiers, including PA (public address) types. When used in Hi Fi-type amplifiers, a small amount of negative (degenerative) feedback is usually fed through a capacitor from the transistor's collector to base.

THE GROUNDED-COLLECTOR AMPLIFIER

Widely used as a unity-gain, impedance-matching stage, the grounded-collector amplifier functions as an electron - tube cathode-follower circuit, most often connected as an emitter-follower. A simplified basic circuit, using an NPN transistor, is shown in Fig. 6-3. The grounded-collector amplifier has a high-impedance input and low-impedance output. It is frequently used—and serves well—as an impedance-matching stage between a crystal phono pickup output and the input of a grounded-emitter stage in preamps. It can even be used to drive a speaker direct, eliminating an output transformer. Notice that the collector is common to both the input and output circuits. Connected as emitter-follower, the circuit is also seen in some Hi Fi equipment power supplies (usually in conjunction with a zener diode), as a voltage regulator.

In audio amplifier practice our common-collector circuit is most often arranged to have a "bootstrap" input, employing a large-capacity electrolytic, to further raise the circuit's input impedance. The precise arrangement, of course, depends on the output impedances of the various components preceding a preamp (phono cartridge, AM/FM tuner, tape player, etc.). Additionally, a bootstrapped grounded-collector circuit is sometimes used between the output of tone-control networks (treble, bass, timbre, loudness) and power output driver stages. Practical bootstrap circuits also can be used in grounded-emitter amplifier stages to increase the input impedance.

TYPICAL PREAMP

The primary purposes of a preamp are (1) to amplify the input signal sufficiently to properly drive the first power amplifier stage, (2) provide necessary input impedance arrangements to match all units to the preamp input (AM/FM tuner, phono cartridge, tape player) and (3) act as a "user control center" for making whatever compensation adjustments the equipment affords.

A grounded-collector, "bootstrapped" preamp using a PNP transistor and connected as an emitter-follower is shown in Fig. 6-4. The 50-mfd electrolytic (C2) is the bootstrap capacitor which increases the input impedance of the emitter-follower stage. In practical use, two of these preamps are directly coupled to the bases of grounded-emitter driver stages in a 25w-per-channel (50w-IHF-Dynamic Power) package-type stereo amplifier.

Considering recording techniques being what they are, and

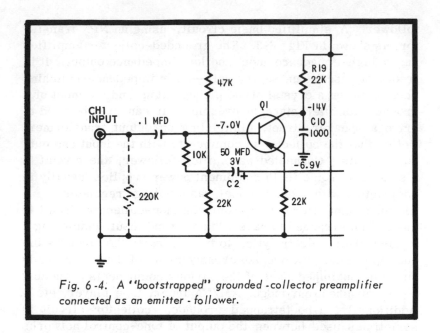

Fig. 6-4. A "bootstrapped" grounded-collector preamplifier connected as an emitter-follower.

hence, the frequency-response characteristics of most material broadcast on radio, recorded on discs, and tape (preemphasis), most quality preamps have some form of compensating circuitry. And we include here all the "owneroperated" controls the equipment contains. This subject is also related to deficiencies or variations in human hearing, especially variations in volume at certain frequencies—usually in the low- and high-frequency areas, plus the problems of variations in "listener-tastes."

Two primary methods, or a combination of both, are employed in the overall process of equalizing, compensating, and stabilizing. These include (1) RC filter networks at the preamp input, between stages or distributed throughout the preamp circuitry and (2) various types of feedback systems —primarily negative (inverse, degenerative) feedback from the output of one or more stages back to the input of one or more preceding stages. Additionally, we are very much concerned with current and voltage feedback, the circuit components and methods employed to stabilize the "Q" or operating point of transistor preamp and amplifier stages.

Some older electron-tube preamps are designed to have a variety of fixed RC equalizing input networks or they may have one phono-type input jack and a switch which allows one or more equalizing networks to be switched into the input circuit. More modern solid-state preamps have multi-wafered switches to provide a number of equalizing arrangements, including frequency-selective feedback circuits.

EQUALIZING NETWORKS

We perhaps remember something about reactive and resistive networks in AC circuits in our previous elementary electronics studies. At least we know from the old formula that capacitive reactance decreases or increases inversely with frequency. And since the circuit load impedance responds in a similar manner, so goes the gain. The arrangement shown in Fig. 6-5A, for example, can be made to attenuate low frequencies over a specified range when properly placed in electron-tube and solid-state preamp circuits. The parallel RC

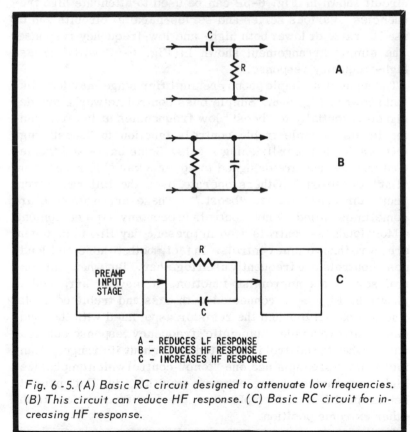

A — REDUCES LF RESPONSE
B — REDUCES HF RESPONSE
C — INCREASES HF RESPONSE

Fig. 6-5. (A) Basic RC circuit designed to attenuate low frequencies. (B) This circuit can reduce HF response. (C) Basic RC circuit for increasing HF response.

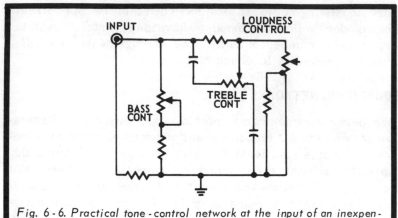

Fig. 6-6. Practical tone-control network at the input of an inexpensive phono amplifier.

circuit shown in Fig. 6-5B can be used to attenuate high frequencies. Various series and series-parallel circuits can be used to raise or lower both high- and low-frequency response. The simple arrangement shown in Fig. 6-5C will increase high-frequency response.

The input of a simple phono preamplifier stage may look like that shown in Fig. 6-6. Simple bass control networks are designed essentially to "boost" low frequencies at lower listening levels. Simple treble controls function to "boost" high frequencies at lower listening levels. Some bass- and treble-control networks are designed to cover a variable range from "flat" to "boost." Others operate over the full range from "cut" through "flat" to "boost." These arrangements are sometimes called "tone" controls in economy-type equipment.

Most loudness controls used in present-day Hi Fi equipment are more than volume controls. In fact, both volume and loudness controls are frequently used together. But when one control serves a compromise function, it may be arranged as shown in Fig. 6-7, connected with bass and treble controls. The two capacitors and the resistor associated with the loudness control provide "automatic" frequency response compensation when the control is adjusted throughout its range. Many inexpensive preamps use one "tone" control which emphasizes high frequencies at one extreme position, is "flat" near its center position, and emphasizes the low frequencies at the other extreme position.

Another arrangement is shown in Fig. 6-8. A two-tap loudness control is used. This circuit, or a variation, is frequently seen in package-type preamps. The value of C1 is selected to provide a measure of high-frequency boost when the control wiper arm is near or below the top tap. C2's value is selected to give similar HF boost over the lower control range. The values of R1 and C3 are selected to provide some bass boost.

We will take a brief look at equalizing and frequency-compensating circuits in one more preamp arrangement and then go on to a more specific thumbnail sketch of the very important subject of feedback. The schematic shown in Fig. 6-9 is that of a preamp having four separate equalized inputs. The relatively high-impedance inputs of the preamp are determined by the circuits which involve the first two transistors. The outstanding feature is a 4-way frequency-selective negative feedback system controlled by switch S1B. Notice the bass and treble controls associated with the last two transistors. These tone control circuits are designed to give full-range adjustment from cut through flat to boost. Notice the feedback arrangements from collector-to-base, associated with the bass and treble controls of the last two transistors. These two feedback arrangements aid in improving the overall tonal response of the preamp.

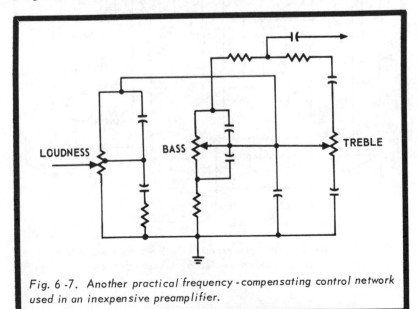

Fig. 6-7. Another practical frequency-compensating control network used in an inexpensive preamplifier.

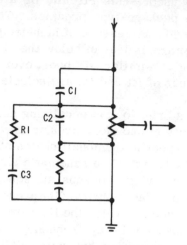

Fig. 6-8. A two-tap loudness control employed in a package-type preamplifier.

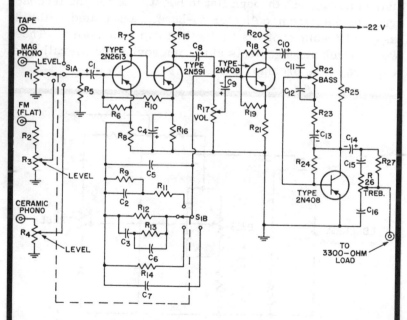

Fig. 6-9. The schematic of a preamp which has four separate equalized inputs, employing selective-feedback compensation. (Courtesy RCA)

FEEDBACK

We cannot place too much emphasis on various feedback methods if we ever expect to approach par for the audio-troubleshooting course. In Chapter 4 it was indicated that an "oscillator is an amplifier is an oscillator." To which we must now add, "an amplifier is an oscillator is an amplifier." Feedback-type oscillators are amplifiers and need positive feedback. Amplifiers need negative feedback and (in most instances) cease to be amplifiers and become oscillators when provided with positive (regenerative) feedback. But most design engineers can prove conclusively that this does not always hold true.

In solid-state amplifiers especially, feedback systems become somewhat more complicated than in electron-tube circuits. And, of course, we cannot cover all feedback ramifications here. But we are compelled to have a cursory understanding of the general principles involved. We have already touched on some feedback principles relating to frequency compensation and solid-state circuit stabilization. Let's expand a little on the subject.

NEGATIVE FEEDBACK

When properly applied in a preamp or amplifier, negative feedback reduces distortion and noise, improves an amplifier's frequency response and stabilizes input-signal amplitude variations. Broadly speaking then, feedback systems are used for stabilizing, limiting, and/or compensating. Each arrangement serves one or more specific purposes. The simple network shown in Fig. 6-10 illustrates how negative

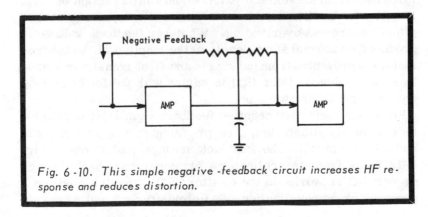

Fig. 6-10. This simple negative-feedback circuit increases HF response and reduces distortion.

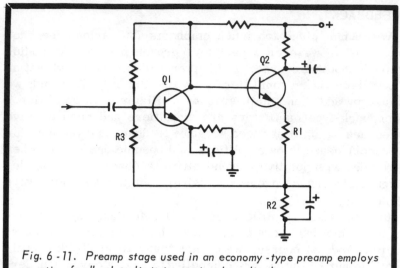

Fig. 6-11. Preamp stage used in an economy-type preamp employs negative feedback to limit input signal amplitude.

feedback is used to provide high-frequency boost and reduce distortion.

Another negative feedback system is sometimes seen in economy-type solid-state preamp stages. Look at the two-NPN transistor preamp stage schematic shown in Fig. 6-11. When a stronger-than-normal positive signal arrives at Q1's base, a negative voltage is developed at the junction of R1 and R2 in Q2's emitter circuit. This 180° out-of-phase voltage is fed back to the first transistor's base—reducing the input voltage. Conversely, when a stronger-than-normal negative signal arrives at Q1's base, a positive-going signal is fed back from the resistor junction and similar action occurs. Negative feedback is used here to "limit" the input-signal amplitude to prevent overloading. Whatever feedback voltage is produced on normal signal levels at the junction of the two resistors is prevented from passing to the first transistor's base by a high-value resistor (R3) in series with the feedback loop and the first transistor's base.

Frequency-selective negative feedback equalizing networks, as previously illustrated, are prevalent in many solid-state and tube preamps. The feedback arrangement shown in Fig. 6-12 feeds from the collector of transistor Q2 back through proper RC networks to the emitter of Q1. Switch position 1 provides phono equalization. Switch positions 2 and 3 provide

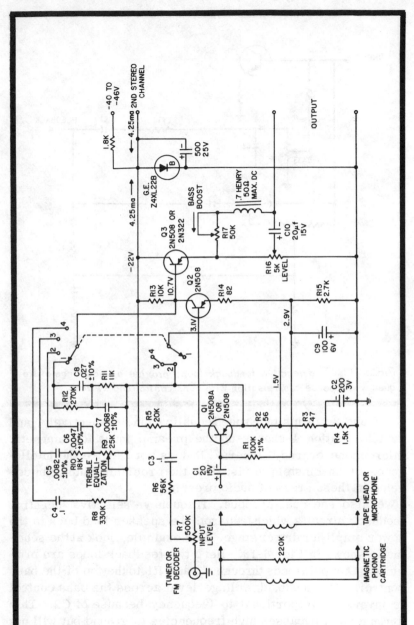

Fig. 6-12. Preamplifier employing four switchable frequency-selective negative feedback compensating networks.(Courtesy General Electric)

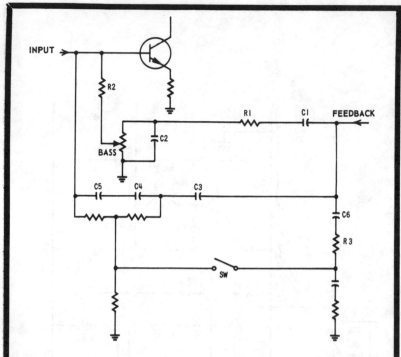

Fig. 6-13. Two negative feedback loops provide a wide range of frequency-response compensation with bass control and switch.

7 1/2 and 3 3/4 ips tape speed equalization, respectively, and switch position 4 shapes-up the preamp input for magnetic microphone or radio tuners. It does not require the intelligence of an Einstein to visualize the problems which can develop in these areas of audio preamps.

We also see dual-purpose, frequency-selective negative feedback circuits which feed from the speaker load back to the power amplifier driver stage. For example, look at the schematic shown in Fig. 6-13. Here two feedback loops are provided. One path feeds through C1 and R1 to the top of the bass control. The feedback voltage level across the bass control is inversely proportional to frequency because of C2. This arrangement bypasses high frequencies to ground but will not interrupt low-frequency feedback—except when the bass control wiper arm is at the ground point. At this point, no low-frequency feedback passes through R2 to the driver transistor base and the amplifier gain is maximum for low frequencies. As the bass control is rotated, more low-frequency feedback is applied to the transistor base. At approximately the control center point, the amplifier's low-frequency response is "flat." Further on, the bass is "cut" as the maximum amount of low-frequency negative feedback is applied to the amplifier's base.

The second feedback path, through C3, C4, and C5, allows only mid- and high-frequency voltage to pass to the driver transistor base. When the switch (SW) is open, the network in this arrangement provides more HF feedback as the frequency increases, hence the driver-stage HF response falls off as the frequencies increase. The switch is closed when the listener wishes to have music at approximately "background" level. C6 and R3 act to increase midrange feedback, reducing midrange frequency response considerably but only slightly affecting the higher frequencies.

CURRENT AND VOLTAGE FEEDBACK

As we have repeatedly indicated previously, you cannot effectively diagnose troubles in preamps and amplifiers, especially solid-state types, unless you understand something about the techniques employed to stabilize the operating or "Q" point. This means stabilizing both current and voltage. Both direct current and voltage feedback systems are employed.

Look at the simplified circuit shown in Fig. 6-14. When DC passes through emitter resistor R1 a voltage drop takes place, making the emitter voltage more negative than the resistor's ground point. Remember, in this circuit (PNP) the base is

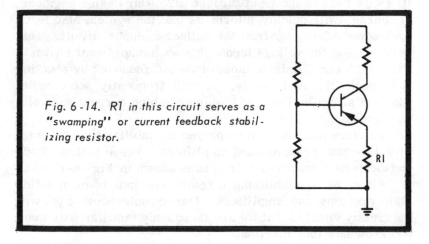

Fig. 6-14. R1 in this circuit serves as a "swamping" or current feedback stabilizing resistor.

R1

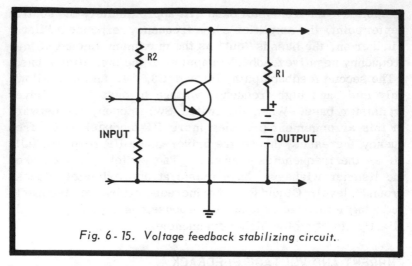

Fig. 6 - 15. Voltage feedback stabilizing circuit.

less positive (more negative) than the emitter by approximately 0. 6v (silicon) or 0. 2v (germanium). If the "Q" point shifts and the collector current begins to increase, in turn increasing the base/collector junction temperature, the current in R1 will increase, decreasing the bias between emitter and base. This in turn decreases collector current, bringing the operating point back to normal. This is a current feedback stabilization system, sometimes called an "emitter swamping" circuit. In practice, the emitter resistor is often bypassed to prevent signal degeneration.

A simple DC voltage feedback stabilizing circuit is shown in Fig. 6-15. Part of the voltage developed across load resistor R1 feeds back to the input through R2. Our design engineer, of course, will quickly inform us that the system also feeds back some AC voltage from the collector output circuit. This is true. But the voltage feedback stabilizing circuit shown in Fig. 6-16 can eliminate most of the AC feedback by shunting it through C1. And finally, we will frequently see circuits like that shown in Fig. 6-17 which combine current and voltage feedback.

Some other methods are employed to stabilize the "Q" point in solid-state preamps and amplifiers. These include diode and thermistor stabilizing circuits as shown in Fig 6-18A and B. All of these stabilizing circuits are important in solid-state preamps and amplifiers. Our troubleshooting job will be greatly simplified if we are thoroughly familiar with them and know how they function.

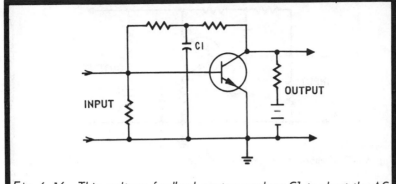

Fig. 6-16. This voltage feedback system employs C1 to shunt the AC component to ground.

BOOTSTRAPPING — POSITIVE FEEDBACK

The word "bootstrapping" is said to have originated with the 18th century "Baron Munchausen" stories written by Rudolph Raspe. The Baron allegedly had the power to virtually lift himself by pulling upward on his bootstraps. "Bootstrapping" has been identified with the damper/flyback "boost" process in TV high-voltage power supplies for more than two decades. It is also identified with some other areas of electronics.

Positive feedback can be used effectively to boost the input impedance of grounded-emitter and grounded-collector amplifiers. Where a higher-than-normal input impedance is required, bootstrapping is especially adaptable to the grounded-collector, emitter-follower circuit. And bootstrapping is a good example of electronic theory working out in practice. The theory makes a preamp input "look," or "appear" higher to a high-impedance transducer than it actually is—logically. But if you argue against the theory, you place yourself in the same untenable position as the pessimist in that optimist/ pessimist story where the optimist sees a light—but there is no light—and the pessimist insists on blowing that light out.

Because the bootstrap circuit is important from the trouble-shooting viewpoint, let's extend our discussion a little further. Look at the grounded-emitter schematic shown in Fig. 6-19. C1 is the "impedance multiplying" capacitor. R1 and R2 are the same, or nearly the same value. An input voltage will divide across R1 and R2, with approximately half of the input voltage appearing at point "A." Roughly, the input impedance

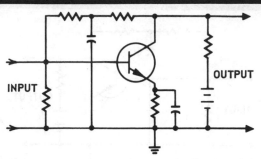

Fig. 6 -17. Combination voltage and current feedback stabilizing circuit.

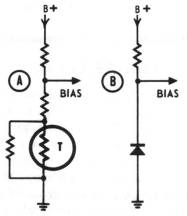

Fig. 6 -18. (A) Bias stabilizing system employing a thermistor. (B) Stabilizing system using a silicon diode.

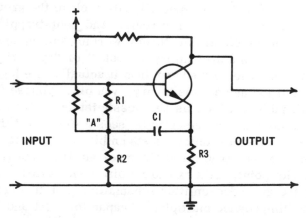

Fig. 6 -19. Grounded - emitter circuit using a "bootstrap" capacitor for boosting input impedance through positive feedback.

is equal to the value of R1 plus R2 (neglecting R3 and the transistor's internal resistance) when C1 is not connected in the circuit. When C1 is in the circuit, it couples an in-phase voltage (positive feedback) from the emitter back to point "A," reducing the voltage drop across R1 and the amount of current through the resistor. It can now be shown (by Ohm's law), that the effective impedance of R1 is increased from 5 to 10 or more times.

"Why doesn't the positive (regenerative) feedback cause the amplifier to "take off" and squeal or motorboat?" we asked. The design engineer we talked to replied, "There's plenty of negative feedback through the unbypassed emitter resistor (R3) to prevent regeneration." Now, if all this makes any sense to you, and you run into trouble with an ailing bootstrapped preamp stage, you should have some idea where the trouble is.

POWER AMPLIFIERS

In our day-to-day audio work we will see power amplifiers connected single-ended and biased Class A, AB and Class B. We will also see similar classed amplifiers connected push-pull. Some other amplifiers will use in-between operating arrangements (AB1 and AB2), but we will not discuss these arrangements. Additionally, other circuits will be found that perform essentially as traditional push-pull amplifiers but use no center-tapped primary output transformer. Finally, we will see more and more complementary-type amplifiers, especially in packaged-type equipment, which employ a PNP and NPN transistor that operate like push-pull amplifiers and likewise use no output transformers.

A simplified push-pull audio amplifier schematic, using PNP transistors, is shown in Fig. 6-20. Each transistor amplifies half of the incoming signal—the positive half going to one transistor and the negative half going to the other transistor from the center-tapped secondary input transformer. Resistors R1 and R2, as we can see, are connected across the power supply and serve as a stabilizing voltage-divider bias network. These resistors are critical. The transistors are also critical and should always be matched pairs.

This circuit is normally biased at Class B, but at a precise operating point (nearer Class AB) to provide a small current flow in each transistor collector at zero-signal input. Other-

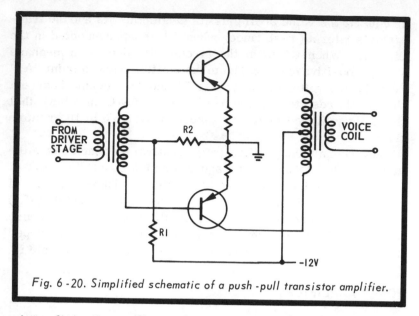

Fig. 6-20. Simplified schematic of a push-pull transistor amplifier.

wise, distortion will result near the crossover point in the swing, especially at low signal inputs. The emitter resistors, as previously explained, provide a measure of operating stability and are also critical in troubleshooting checks. In reference to the simplified schematic, it should be noted that most practical push-pull Class B amplifiers employ negative feedback (from output to input) to further reduce crossover and other types of distortion.

SINGLE - ENDED CLASS B

The effort to lower production costs of low- and medium-power amplifiers has resulted in the design and development of an increasing number of variations in amplifiers, especially package-types. These include the Class B (or AB) group —frequently called "single-ended" or "series-connected" types which operate essentially in the push-pull mode. Some use input transformers having two separate bifilar-wound secondary windings instead of the conventional center-tapped secondary and no output transformer as shown in Fig 6-21. The low-impedance output of these amplifiers may be directly or capacitively coupled to a speaker voice coil.

The two transistors conduct alternately, one transistor amplifies the positive half of the input signal and the other transistor amplifies the negative half. One transistor is always cut off while the other is operating and passing the amplified current to the speaker voice coil. At the crossover (zero) point, where one transistor "fires" and the other cuts off, distortion on weak input signals will result if the firing sequence of the alternate on/off half-cycles are not perfectly timed.

Other amplifiers in this group may be driven directly (without an input transformer) by various means. One direct-drive method employs the so-called "Darlington" Class B push-pull, emitter-follower arrangement as shown in Fig. 6-22 (Q2 and Q3). This arrangement is not to be confused with the "Darlington-Pair," 2N997 device. Transistors Q2 and Q3 drive a single-ended Class B push-pull amplifier (Q4 and Q5).

The complementary-symmetry amplifier is another manifestation of the trend toward lower cost audio equipment. Here, a matched pair (having similar characteristics and ratings) of NPN and PNP transistors are used. The opposite-polarity transistors are usually driven in parallel while their outputs are connected through a common lead. Both input and output transformers can be eliminated and still maintain some of the advantages inherent in push-pull operation. A basic, simpli-

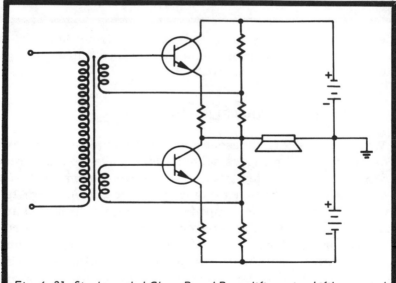

Fig. 6-21. Single-ended Class B or AB amplifier using bifilar-wound input transformer secondaries. These secondaries are marked for proper polarization. The speaker is directly connected.

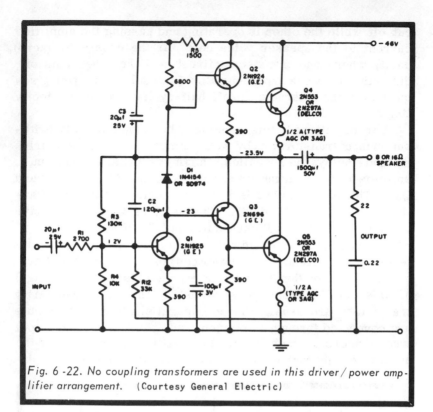

Fig. 6 -22. No coupling transformers are used in this driver / power amplifier arrangement. (Courtesy General Electric)

fied complementary-symmetry amplifier circuit is shown in Fig. 6-23 where RL represents the speaker voice coil. A practical circuit is shown in Fig. 6-24.

MODERN HI FI STEREO AMPLIFIERS

What is characteristic about many of the amplifiers we have described in the previous section which make them different from those used in top-grade Hi Fi stereo equipment? Not very much, but there are some differences. Although the basic mono preamps and amplifiers previously discussed are similar to many used in stereo equipment, we must now re-orientate our thinking somewhat beyond the common viewpoint that "a stereo amplifier is just two mono amplifiers working together simultaneously." This is necessary for a number of legitimate reasons.

A very large percentage of the home audio equipment made today is stereo-structured. And our basic approach to stereo preamps and amplifiers is different: Because there are two amplifiers and both seldom become defective at the same time, our approach is to use the good amplifier to help us locate trouble in the defective unit. To a great degree this makes our work easier, but we must know the amplifier circuits, have complete manufacturer's service data and use the proper test instruments effectively, otherwise stereo amplifiers can be difficult.

From the maintenance viewpoint then, stereo preamps and amplifiers designed today and those that will be designed tomorrow, are now and will be somewhat more sophisticated

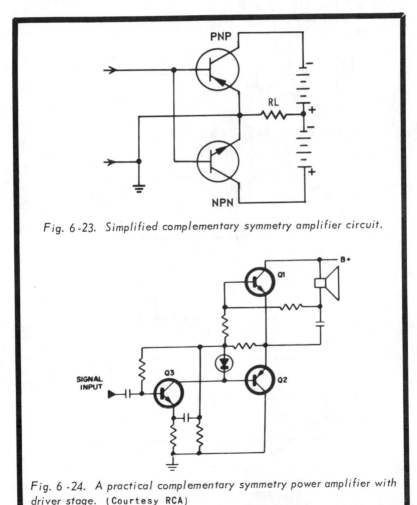

Fig. 6-23. Simplified complementary symmetry amplifier circuit.

Fig. 6-24. A practical complementary symmetry power amplifier with driver stage. (Courtesy RCA)

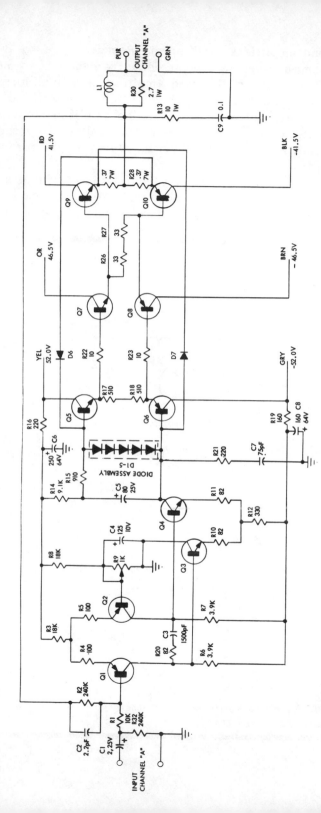

Fig. 6-25. Schematic of one channel of the J. B. Lansing SA 600 and SA 660 stereo preamp / amplifier.

than those we have previously reviewed here. This is not to imply that we can expect a sudden rash of exotic amplifiers —products of buzz-worked breakthroughs. Market research stands too staunchly between product research, development, and the assembly line. But slow evolution has a way of sneaking up on us.

The amplifier in Fig. 6-25, although a few years old, is representative of a high-class amplifier. A schematic of one channel of the J.B. Lansing SA600 and SA660 amplifier is shown in Fig. 6-25. This amplifier has two basic sections: The driver section, consisting of four transistors, is a two-stage differential (difference) amplifier. The output section, consisting of six transistors, is arranged as three cascaded complementary-symmetry emitter-follower stages. In effect, three transistors function on positive half-cycles and the other three conduct on negative half-cycles. The five diodes supply stable bias to the six final amplifier transistors.

Notice the following characteristics of this amplifier: (1) All stages are direct-coupled throughout and no transformers or coupling capacitors are used (except one coupling capacitor at the driver input). (2) The feedback loop goes from the output all the way to the driver input. (3) The symmetrical arrangement of the output circuit makes it almost immune from power-line surges. (4) Successively lower collector voltages are used from the final amplifier input to the output. (5) Each pair of transistors in the output stage has a higher beta cutoff frequency than the following pair. (6) The amplifier has

Fig. 6 -26. Front view of the JBL SA 600 and 660 preamp /amplifier.

Fig. 6-27. Electro-Voice 1244 preamp and amplifier.

a thermal overload breaker (not shown in schematic) to guard against a short circuit. A photograph of the SA660 preamp/amplifier control unit is shown in Fig. 6-26. Another solid-state preamp/amplifier control unit is shown in Fig. 6-27.

STEREO AMPLIFIER ARRANGEMENTS

In addition to the "tone" controls (bass and treble), the volume or loudness controls used in modern preamps—adjustments that allow the user to emphasize or diminish low and high frequencies and vary the output level—the stereo preamp has a balance control which is not used on mono preamps. Each preamp has one each of the aforementioned controls (but ganged) to be operated by one knob on the preamp. When the balance control is adjusted at its mid or zero point, it provides a so-called "three-dimentional" audio product from two properly-spaced speakers. When this control is moved in one direction, the product from one speaker is emphasized and when moved in the other direction, the product from the other speaker is emphasized. Normally, the control is set at zero, although acoustic conditions in the listening area may require the control to be set to the right or left of zero for best listening results.

Some modern stereo preamps have more versatile volume/loudness control arrangements. For example, the basic arrangement may have a straight-line volume control which can be adjusted from silence, to a whisper, to a roar. In other words, the volume control is strictly a volume control. A "loudness" switch may be employed in conjunction with the volume control to switch in a network which emphasizes both low and very high frequencies when the volume control is adjusted downward.

As we have already said, transistors are more critical than tubes because of their characteristics. The design steps that must be taken to limit "Q"-point variations, for example, result in the addition of critical circuitry that refuses to tolerate slipshod approximations. And modern, high-class stereo preamps and amplifiers take full advantage of these circuits. This includes bias and bias-stabilizing resistors, feedback-circuit resistors and capacitors, stabilizing diodes, current-limiting and protective circuits, diodes and thermistors for voltage, current and temperature compensation. These are all critical service areas.

Most amplifiers designed today have little lee-way for plus or minus component variations. Most conventional Class B and single-ended push-pull-type Class B amplifiers used in portable and many package- and integrated-component type stereo amplifiers, are actually biased Class AB to reduce cross-over distortion which occurs near the zero-swing point as shown in Fig. 6-28. This means that a small amount of current must flow in the collector circuit under no-signal conditions. Even so, with very small input signals some distortion still remains—necessitating a modest (sometimes husky) amount of negative feedback to suppress it further. Cross-over distortion is caused by a particular characteristic of transistors, a low forward-current transfer ratio at low

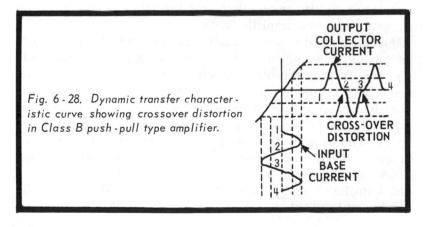

Fig. 6-28. Dynamic transfer character-istic curve showing crossover distortion in Class B push-pull type amplifier.

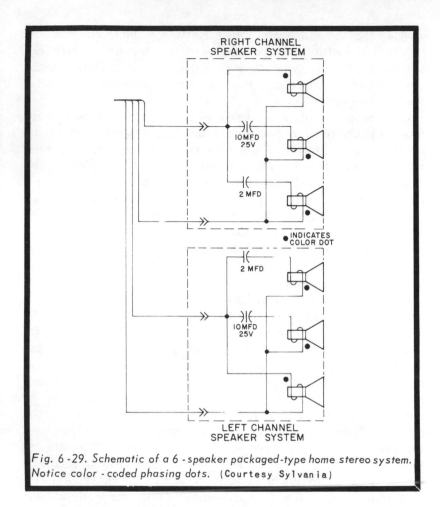

Fig. 6-29. Schematic of a 6-speaker packaged-type home stereo system. Notice color-coded phasing dots. (Courtesy Sylvania)

currents. But it can be eliminated for all practical purposes in better designed amplifiers.

Most integrated preamp/amplifiers have various front-panel switches, including on/mute for speakers (when using headphones), loudness control on/off switch (some audiophiles don't like loudness controls), selective-feedback switches for equalized phono, tuner and auxiliary switches for tape-monitoring, etc. Other preamp/amplifier units may have up to a dozen or more switches. And the speaker arrangements used in stereo installations are somewhat more complicated than in mono arrangements as shown in Fig. 6-29. The preamp and amplifier stages of the Motorola FM990X FM/stereo radio mentioned in Chapter 5 are shown in Fig. 6-30 (A) and (B).

TROUBLESHOOTING MODERN AUDIO AMPLIFIERS

The slightest noise originating in a preamp may sound like a disintegrating pack of firecrackers in the speaker. The slightest hum under similar conditions becomes an old-style foghorn with its switching mechanism stuck at the on position. While somewhat exaggerated, these observations are made for the purpose of directing your attention to the key problem in stereo preamps and setting the stage for effective troubleshooting techniques. The approach is similar for both electron-tube and solid-state preamps and amplifiers.

Although today's Hi Fi service problems have not changed basically in the past two decades, the problems have multiplied and they are now more sharply defined. This is true for two primary reasons: (1) Equipment specifications have become more highly refined and (2) The listening tastes of the general public have become more discriminating. Few music lovers, for instance, would bother to listen to the output of a 20-year-old Hi Fi installation. When today's listener has a Hi Fi preamp or amplifier repaired, it must sound as well afterward as it did before. The major problem and its solution, then, seems obvious: the knowledge, the service data, the test instruments must be at hand to make effective repairs. And after the repair is made the equipment must be checked out to make sure it conforms to the manufacturers' original specifications.

GENERAL SERVICE PROCEDURE

The trouble symptoms in preamps and amplifiers are about the same as those listed in Chapter 3 for AM radios, but the number of causes for the symptoms are somewhat less. There's no point in repeating the list. And the modulated signal-injection technique, employing a speaker at the unit's output, is used to isolate some defects. But in many cases, we will need all the instruments mentioned in Chapter 2.

In most cases, however, it is easier to isolate and pinpoint faults in preamps and amplifiers than it is in AM radios. This seems obvious since we are concerned primarily with easy-to-trace audio-signal stages and have no RF, oscillator, mixer, IF or detector stages to worry about. It is also easier to isolate and pinpoint trouble in single unit (separate preamps and amplifiers) than it is in those integrated units where pre-

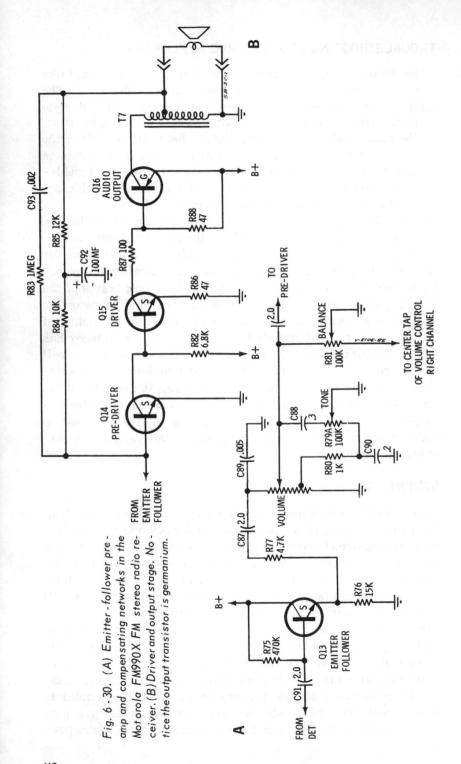

Fig. 6-30. (A) Emitter-follower pre-amp and compensating networks in the Motorola FM990X FM stereo radio receiver. (B) Driver and output stage. Notice the output transistor is germanium.

amp, amplifier, and power supply are all mounted on one chassis. This also is obvious, since it is easy to substitute preamps and amplifiers with known-good units to initially isolate certain faults to one or the other. But stereo amplifiers, as previously mentioned, are even easier if we have a dual-trace scope or a good flip-flop.

Most troubles in electron-tube preamps and amplifiers originate in defective tubes. Always substitute tubes with one or more known-good ones—do not lean too heavily on the results of tube testing. When you replace tubes or transistors in conventional push-pull, 2-transistor single-ended Class B or complementary-symmetry type amplifiers, check the current balance in the grids (or bases) as described in Chapter 3 under the section "Distortion—Normal Volume." Both tube and transistor pairs in push-pull type amplifiers must be matched—have similar characteristics—and operate under balanced drive, bias, and load conditions. Otherwise, you'll have a distortion problem. Under severe unbalanced conditions, you'll also have short-lived tubes and transistors because of overloading in one half of the circuit.

Once again, on "dead" preamps and amplifiers, first check fuses or other power supply safety devices, power supply output voltage, on/off switch. Then inject a test signal to check the speaker, output tubes, output transformer (if any), coupling capacitor (if any), and then stage-by-stage toward the front end. When the defective stage is isolated, make voltage and perhaps resistance checks. It's straight-forward trouble-shooting—no big deal.

But suppose you have no normal output from the speaker and you get a hum when the loudness or "volume" control is adjusted full up? Look for a defective driver or preamplifier stage. Signal-inject to isolate the stage and check voltages, substitute the tube, check the transistor. If the voltage is off more than 10 or 15% at any tube or transistor element, check all resistors and capacitors in the stage.

HUM AND INTERFERENCE PROBLEMS

"Hum" is a general problem with preamps and amplifiers. Under certain conditions it can become a "dog" job to locate the source. This is especially true with older electron-tube preamps and amplifiers. But many hum problems have disappeared in modern solid-state preamps and amplifiers, especially the problem of ground-loop hum.

Most hum will be 60 or 120 Hz (or higher harmonics of these frequencies). In some older electron-tube amplifiers you may need to adjust a "hum control" to balance the hum out. In tube amplifiers where heaters operate from AC, the equipment may develop a heater-cathode short in one tube which will provide a husky 60-Hz hum. A defective power supply filter in a full-wave power supply may cause heavy 120-Hz hum. A scope is a fine instrument for isolating the source of hum. In marginal cases, a calibrated, sensitive scope can be used to measure the percentage of power supply hum or AC ripple.

An open grid resistor can cause hum. You can tell if the resistor is open when you measure the grid voltage with a VOM —the hum will usually decrease, sometimes to a very low level, while the meter is in the circuit. In some older pushpull power amplifiers, hum can result from unbalance caused by off-value grid resistors.

Microphonic tubes can cause a heavy "howl" in the speaker. Tap the tubes lightly at different volume levels (but do not let the speaker howl at a high wattage level—it or the output transformer may be damaged). Replace the microphonic tube when you locate it, of course.

As the demands of the nation's electronic equipment needs increase, we will be confronted by more "interference" problems. For example, nearby high-powered ICBM radar missile-detecting equipment can, and does, cause unusual problems for all kinds of home- and commercial-audio communications equipment. In fact, we have experienced cases of hum, buzz, and distortion in Hi Fi-type audio amplifiers from both fixed and mobile transmitters operating in the public, commercial, military, law-enforcement and amateur radio services.

It has been indicated previously how an amplifier can become an oscillator under certain conditions. But a highly sensitive preamp or amplifier can also become a receiver by rectifying and detecting RF signals in its input circuit. A strong, nearby RF signal can cause an amplifier to overload, increasing the grid bias to a point where rectification results through clipping action. This usually means hum, buzz, or distortion. The effect may be steady or intermittent, depending on the type and nature of the RF transmission.

You can isolate a detecting electron-tube type amplifier stage by using a calibrated scope and by grounding tube grids momentarily. Don't waste your time trying to determine the interferring signal's frequency. In this situation, it makes little difference what the frequency is.

Watch for unshielded tubes, unshielded input leads and high-impedance grid-bias circuits. You may be able to solve a problem of this type in Hi Fi preamp and amplifier stages by (1) completely shielding the entire amplifier or by (2) installing a shielded RF choke having an inductance value between 0.5 to 1 mh in the tube's control grid directly at the socket terminal.

Strong RF can cause various types of interference also in sensitive solid-state amplifiers, depending on the design and methods employed to limit input signal strength and stabilize bias conditions.

MANUFACTURERS' SERVICE AIDS

As previously mentioned in Chapter 2, step-by-step procedures for measuring preamp and amplifier distortion are outlined in operating manuals which come with sine/square wave generators, distortion meters, and other test instruments. But, equally important, it is well to remember that most manufacturers provide, for a small charge, service data and bulletins covering specific repair problems concerning their equipment. The initial cost of this data is insignificant compared to the ultimate profit you make through time saved and by providing more satisfactory service to your customers.

Manufacturers' "Production Run Changes" are only one of these important service aids. In addition to specific directions for solving distortion and other problems (by substituting different-value resistors and capacitors primarily), these bulletins describe actual in-field cases, how they were solved, and list many practical time-saving tips.

For example, we ran into a problem with a medium-power solid-state amplifier some time ago with a case of severe distortion. The owner claimed it had been repaired three times in one year for the "same complaint, and hadn't sounded right since the first repair." We asked the owner a lot of questions about previous repairs. How did the amplifier sound before the first repair? Did he know what actual work had been done on the amplifier? Were the repairs made in the home or in the shop? Among other important things, we discovered that three power-output transistors had been replaced at various times within a year. We took the ailing amplifier to the shop and started looking up the manufacturer's service data.

We already knew, of course, that a mistake had been made by changing one transistor, instead of both with a matched pair, but we wanted to learn why a transistor failed in the first place. We found three pages of information in our files from the amplifier manufacturer. This information not only helped us to properly correct the distortion problem in short order, but we also discovered why the first transistor failed and how to prevent the same thing happening again.

When one transistor fails in push-pull or in push-pull type amplifiers (this applies to similar-type driver stages also), precautions must be taken:

(1) Check all resistors and capacitors in the stage.

(2) Replace both stabilizing resistors with a matched pair. And if these resistors are small values, make sure they have 5% (gold stripe) tolerance ratings. (NOTE: Some solid-state output amplifiers may use fuses in emitter circuits. A 1/2A 3AG or AGC fuse, for example, has a resistance of about 1 ohm and may serve the dual role as protective device and emitter-stabilizing resistor.)

(3) Remove both old transistors, and with normal power applied to the amplifier make measurements at the transistor sockets for base, emitter, and collector voltages. If voltages are not normal and equal at similar element points, look for an off-value resistor or a leaking capacitor.

(4) When you are sure everything is in order, switch off the amplifier power, install the new transistors and give the amplifier a heat run.

It is surprising how many technicians (especially those who have been in the business for the past two decades or so), disregard equipment pin-jack and other type test points which manufacturers provide on circuit boards and clearly indicate on schematics and component board layouts. These have been provided, at considerable cost, to help speed your trouble-shooting and repair work. Use them—and prosper.

INTEGRATED CIRCUITS

We briefly mentioned integrated circuits (ICs) in Chapter 5. Some trends indicate that microelectronic, or integrated circuits, in one form or another—perhaps in a variety of forms —will have more impact on home- and commercial-audio equipment in coming years than the transistor has to date.

But here, as in other areas of this book, we are concerned only with the practical aspects of the subject from a maintenance viewpoint. It would not serve our interests to fully explore "microelectronic" or "modular" electronic technology. We do not care if the circuits are "hybrid" (multichip) or "monolithic." In fact, we are not even concerned with the major theoretical concepts of "differential" (difference) and "operational" amplifiers nor with digital IC circuits at this point. When digital-type circuits begin appearing in home- and commercial-audio equipment we'll be advised about it somewhat in advance.

We are interested, however, in certain aspects of linear-type ICs even though it may be a long time before we see ICs having sufficiently large resistive and capacitive (to say nothing of inductive) components to replace all "discrete" components in audio equipment. It would seem reasonable to assume also that we are now seriously interested in those single-packaged circuits which contain relatively small-value passive (resistors and capacitors) and active components (transistors and diodes). At least, the present trend in modular-designed home and commercial audio equipment indicates that we should be concerned.

If this trend continues along the same lines, then trouble-shooting and repair will consist of a simple two-step process (1) isolate the defective module and (2) remove it and plug- or solder-in a new off-the-shelf module. The old module will either be returned to the factory (or distributor) or thrown away. But most manufacturers are insisting that an IC in their equipment is the "least likely part to fail."

Another approach closely related to this trend is the "substrate amplifier" design introduced by RCA about a year ago. Here, most components are contained on a ceramic wafer

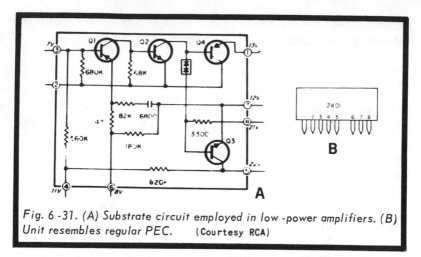

Fig. 6 -31. (A) Substrate circuit employed in low -power amplifiers. (B) Unit resembles regular PEC. (Courtesy RCA)

somewhat similar to a PEC. This circuit, called the Z401, is shown in Fig. 6-31A. Its 8-connection packaged profile is shown in Fig. 6-31B. These were used primarily in por- table phono amplifiers but no doubt have by now found their way into other equipment. A number of other ICs have been employed in audio equipment during the past few years. A number of higher-powered complete IC amplifiers have already been designed.

One important thing to look for in ICs are manufacturers' production changes and service aids. For example, in one case when a defective IC circuit was replaced in a phono am- plifier, the amplifier developed both a low (30 Hz) sine-wave hum (detectable only with a scope) and a higher frequency saw- tooth buzz. In this case, the manufacturer's service infor- mation advised what changes were necessary to solve the prob- lem. Most manufacturers provide ample service instructions for isolating IC faults to either the IC or to external discrete components.

AVOIDING MORE TROUBLE

If we repeat here, it is only because years of practical ex- perience show that some things bear repeating. Many prob- lems arise and many "profit-eating" call-backs occur because some very important areas of electronic theory are disre- garded, overlooked, or not practiced. In servicing solid- state preamps and amplifiers particularly, we cannot disre- gard this theory without paying penalties.

When a power transistor fails, the chances are it does so for reasons which exist external to the transistor. Few transistors fail because of a manufacturing defect in the transistor. When they do fail, try to find out why and don't replace them until you locate the external fault. Always make resistance checks in the external circuits before replacing transistors. Make sure no fault exists before a new transistor "blows" also.

Once again, don't replace one transistor in a push-pull output or other balanced stage. Replace both with a matched pair (similar alpha and beta)—but only after you know that no circuit overload problems exist.

Do not "jump" electrolytics in solid-state preamps and amplifiers while the equipment is operating. Disconnect one end of the suspected component, "tack in" the new one, and then switch the equipment back on.

Do not operate an amplifier with the speaker disconnected. Place the proper load across the amplifier's speaker terminals —and do this while the equipment is switched off. Incidentally, on some solid-state amplifiers you can "blow" the output transistors by operating the amplifier with the incorrect speaker.

Do not replace power supply diodes without checking for shorted or leaking power supply or other filters.

Some manufacturers recommend that a thin film of silicon grease (DOW Corning DC4) be applied to both sides of the mica mounting insulators and also from heat sink to chassis. Instructions should be followed. In some cases heat sinks are made rather large and are insulated from the chassis; the heat sinks operating at collector potential. Where silicone grease is used between collector and heat sink, no mica insulation is employed.

Solid-state circuitry is critical because of the nature of transistors. When you replace a resistor or capacitor, make sure values are correct. For example, you can make yourself an expensive callback by placing an off-value capacitor or resistor in tone control circuits. The same with an incorrect value coupling capacitor. In stereo equipment, particularly, this can throw off amplifier balance and most Hi Fi equipment owners will detect the difference and complain. And finally, some tape recorder preamps use special low-noise type resistors, having 5 or 10% tolerance.

CHAPTER 7

'Combinations' and Related Problems

Most multiband and combination radio receivers appear complicated because we allow ourselves to incorrectly assume that they are. If we analyze the situation briefly, this fact becomes immediately obvious. Study a few combination schematics and you will see, once again, that we must think of various radio combinations as "whole systems" made up of separate parts. A combination AM/FM radio, for example, is simply another type radio with two separate functions. But one important difference exists: Some components in the whole radio function on AM, some on FM, and some function in both modes. And small AM/FM combinations are, for all practical purposes, almost identical to many types used in consoles except the power output in consoles is frequently somewhat higher and the function switch is a three-, four-, or five-way type for operating a phonograph, tape recorder, or TV, in addition to AM/FM/stereo.

SIMPLE COMBINATION

If we observe a typical portable or table-top transistorized AM/FM mono radio, we find that it may employ 10 transistors and eight diodes. Depending on whether the AM/FM function switch is in the AM or FM position, the transistors may serve as follows:

The first transistor may serve as an untuned FM RF amplifier, the second as an FM converter, the third as an AM converter. But the fourth transistor may serve in a dual role as a first AM and FM IF amplifier, the fifth as second AM/FM IF amplifier, the sixth transistor may operate only as a third

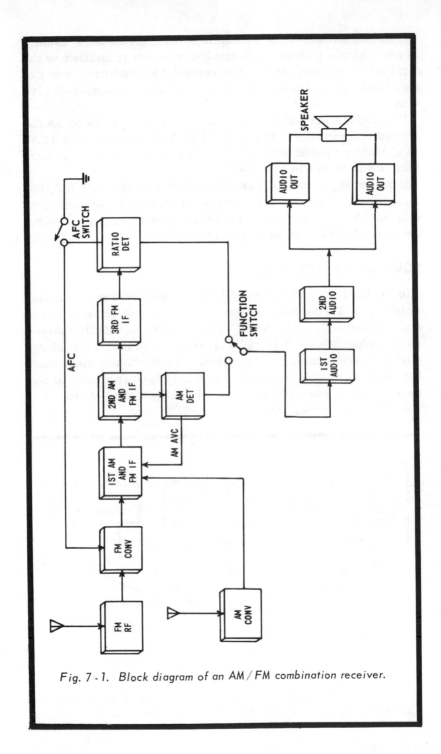

Fig. 7-1. Block diagram of an AM/FM combination receiver.

FM IF amplifier. The remaining four transistors usually serve a single role as audio amplifiers—one transistor as the first AF amplifier, one as the second AF amplifier, and two as a matched audio-output pair in a Class B, push-pull circuit.

Two diodes are used in a ratio detector, one diode as the second AM detector, one in the FM AFC circuit, one as an AGC rectifier, and three as AGC "overload" diodes. A block diagram is shown in Fig. 7-1.

Some package combination front ends use separate AM/FM RF stages, separate oscillators and mixers (or converter stages), and a common three-IF section throughout. Some use two AM IFs and three FM IFs.

COMBINATION SWITCHING

In smaller combinations, like the AM/FM radio just described, a 3-position, 3-wafer function switch is provided and marked AM, FM, AFC. This switch is usually marked on the schematic—SW1 or SW2A, SW2B, and SW2C and shown with the various circuit connections. As previously mentioned, the function switch in combinations may have from two to five or even more positions and a number of wafers—up to 10 or more in some combinations.

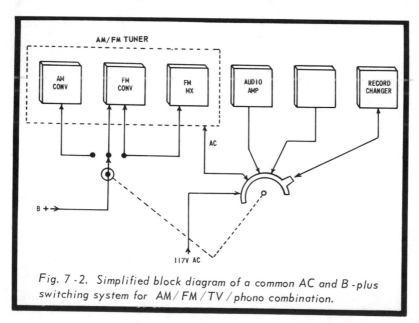

Fig. 7-2. Simplified block diagram of a common AC and B-plus switching system for AM/FM/TV/phono combination.

In some package home combinations, the AC inputs to the various units are common, as well as the audio output and speaker systems. A simplified block diagram of a common AC supply in a 5 - function arrangement is shown in Fig. 7-2. In this system a second function-switch wafer supplies B-plus to the AM/FM tuner from a power supply located on the amplifier chassis.

Function switches on some equipment become rather elaborate. For example, a multi - function, ganged wafer switch arrangement may include switch points for AM, FM, FM/stereo, phono - mono, phono - stereo, and other functions. Also, a switch is usually provided to disable the FM oscillator AFC when an FM signal is not being received.

In some combinations that contain AM/FM tuners, automatic record changers, and TVs, the TV has its own individual audio output which is switched to common speakers in the console cabinet. In other combinations the TV's sound detector feeds into a common audio driver and output system which serves the AM, FM, phono, or tape player. Unfortunately, for technicians, few industry standards prevail in combo sets. But this problem need not pose unusual service difficulties for well-trained technicians and experienced service-dealers. AM and FM tuners, TVs and phonographs, tape recorders and players—no matter who makes them—are basically similar, whether mounted singly or in combinations. And schematics of the various combination arrangements are available from manufacturers. The most important consideration in handling combinations is the service approach. Applied to combinations, this means knowing that the function switch is the key to breaking down the whole combination into separate functional parts. The switch automatically isolates the faulty section.

WORKING WITH COMBINATIONS

The troubleshooting techniques which apply to all kinds of combinations are well known to skilled technicians. The initial approach—the modus operandi—is simple: Check all function-switch positions. If the equipment does not work on any function position, it is then clear that we must look for trouble in those areas common to the entire combination. What circuits are common is best determined by a quick glance at

Fig. 7-3. Lear Jet table-top or bookshelf integrated-component stereo music center.

the manufacturer's schematic. But in the case mentioned here, we know instinctively that it's either a blown fuse or other protective device failure affecting the power supply, the power output stage, or speakers. Early in this situation we check for B-plus at the power supply output. It could also be a tube or transistor or any component common to all functions. This holds true for all combo equipment, including AM/FM/SW/phono/tape-recorder/tape-player units.

If the equipment works in one mode and not in the other, do not look for trouble in the common circuits. We obviously check only those circuits and components associated with the defective operating mode. And at this point it should be remembered that switch contacts sometimes become defective —permanently or intermittently.

In many package-combinations, when the phonograph and

Fig. 7-4. H. H. Scott packaged-type console music center.

FM outputs are normal and there's no AM output, check the AM converter tube or transistor stage early in the trouble-shooting procedure. If the trouble is not here, check the function switch continuity with a signal from a modulated signal generator or noise generator. In many cases this is done by applying the signal to the AM detector output point. The procedures are not unlike those already described for trouble-shooting AM and FM radios. We need only check those components common to a particular function.

The HI FI combinations you will be called upon to service will range from table-top "compacts" to table-top or book-shelf type integrated-component types to a wide selection of elaborate consoles designed to mix with any type furniture or room decor.

For example, the equipment shown in Fig. 7-3, an integrated-component type music system provides an AM tuner, FM stereo tuner, a 4-speed automatic record changer, automatic 8-track stereo tape cartridge player, stereo preamps, amplifiers and two speakers. The console type combination shown in Fig. 7-4 provides the same facilities without tape player which can easily be installed as an external accessory unit. Some package-type consoles have space to install tape players internally.

CHAPTER 8

Solid-State Power Supplies

Tighter power demands of solid-state home- and commercial-audio equipment have brought about more sophisticated power supplies. Voltages required are small and the operating range of transistors is very narrow. And transistors, voltage-dependent resistors, PN junction diodes, including the zener, have combined their individual characteristics to make adequate power supplies possible.

Disregarding battery-powered equipment, virtually all transistorized AM, FM, AM/FM, AM/FM/SW and other combinations—including stereo preamps, amplifiers, tape recorders, tape players, and portable phonographs now have solid-state power supplies. And many of these supplies are provided with one or more voltage-stabilized outputs and current-limiting protective arrangements.

HALFWAVE RECTIFIERS

First it is appropriate to warn you not to be confused by the plus sign that some manufacturers place on the cathode end of a silicon rectifier or next to the cathode on schematics. But when you replace one, be sure it is polarized in the circuit exactly like the original was. If not, the current phase will be reversed and produce some undesirable results. Aside from this, electron current flows through a diode from the cathode end—opposite the direction of the arrow-headed anode end as indicated by the left-curving arrow shown in Fig. 8-1. For each one cycle of AC input to this circuit, you get one positive half-cycle out. Since this diode, as connected, is forward-biased only on positive half-cycles, it will not con-

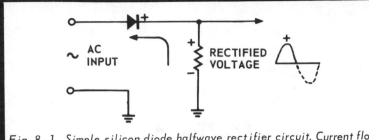

Fig. 8-1. Simple silicon diode halfwave rectifier circuit. Current flows from the cathode to the anode, around through the load.

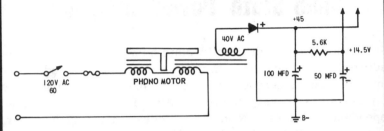

Fig. 8-2. Halfwave rectifier circuit used in a phonograph. Part of the power transformer consists of the phono motor windings.

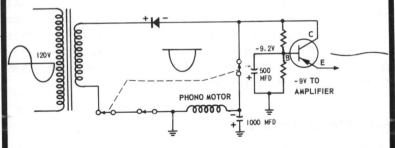

Fig. 8-3. Negative-going halfwave power supply having capacitive multiplier transistor to stabilize voltage and filter out ripple.

duct during negative half cycles. Hence, only the positive half-cycles appear at regular intervals in the output. This is a halfwave rectifier, and when provided with proper filter capacitors, it makes a reasonably good AC power supply for certain applications.

Should you reverse this diode, the output phase will shift 180° (reverse) and the ground point of the load resistor would become positive and the high end would become negative. In this simple, isolated circuit no "fireworks" would result. But in many power supplies, things would happen. As we shall see shortly, however, some power supplies do use one or more silicon diodes in the "reverse" direction.

The halfwave rectifier shown in Fig. 8-2 is used in a solid-state "portable" phonograph. Notice the motor windings are an integral part of the power transformer. The halfwave circuit shown in Fig. 8-3 has a "reversed" silicon diode in the circuit to provide a negative supply. The transistor base is supplied a stable voltage by action of the 500-mfd electrolytic and the 2-resistor voltage divider. This stable bias arrangement between the emitter and base results in a stabilized -9v output. The two large filter capacitors, together with the transistor action, provide an almost pure DC output having a low ripple content as shown.

A simplified fullwave rectifier circuit is shown in Fig. 8-4. A center-tapped transformer and two diodes are used. In effect, this arrangement serves the same as two halfwave transformers in series. The center tap of the transformer secondary is always positive to the nonconducting diode and negative to the conducting diode. On the positive half-cycle as shown, D1 conducts and current flows through the top half of the transformer, through the load and back through D1. Diode D2 is back-biased and cannot conduct. On the negative half cycle, diode D2 conducts and current flows through the bottom half of the secondary, through the load and back through D2. D1 is back-biased and cannot conduct. Another fullwave rectifier (bridge) employs four diodes and, like a halfwave rectifier, does not require a transformer unless a voltage output higher than the input is needed. And a center-tapped transformer is not required in this arrangement. The opera-

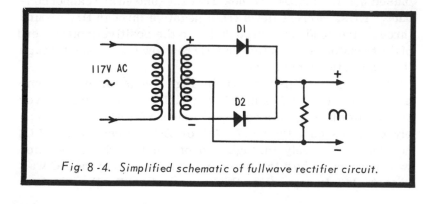

Fig. 8-4. Simplified schematic of fullwave rectifier circuit.

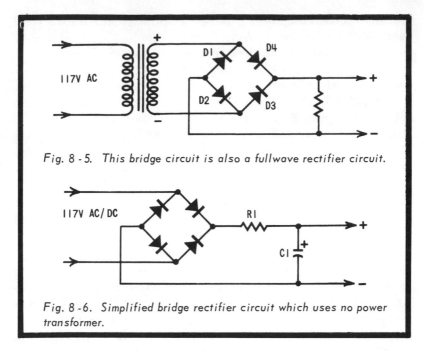

Fig. 8 -5. *This bridge circuit is also a fullwave rectifier circuit.*

Fig. 8 -6. *Simplified bridge rectifier circuit which uses no power transformer.*

tion of this circuit is a little tricky, but a second glance will show how it works (see Fig. 8-5).

We can see at first glance that D4 is forward biased with a positive pulse on its anode. And D2 is also forward biased with a negative pulse on its cathode. But D1 and D3 are reverse-biased and cannot conduct. Thus, current flows from negative through D2, around through the load, through D4 to the plus (top) end of the transformer. When the top end of the transformer becomes negative and the bottom end becomes positive, we can again see that both D1 and D3 are forward biased and D2 and D4 are now reverse-biased and cannot conduct. Thus, current flows from negative through D1, around through the load and through D3 to the positive (bottom) end of the transformer. Notice that current always passes through the load in the same direction.

The bridge rectifier shown in Fig. 8-6 is used in some medium-powered PA amplifiers which employ silicon power transistors having high voltage-breakdown ratings. The supply operates on either 117v AC or DC. When used on DC, current passes only through two of the four diodes—which two depends on how the line cord is plugged into the DC wall receptacle. In any event, the bridge automatically polarizes the DC voltage at the power supply output. R1 is wire-wound (WW), usually 10 ohms 20w in amplifiers rated at 20 to 25w. C1 is usually around 200 to 300 mfd at 150v.

SPLIT - VOLTAGE ARRANGEMENTS

In some solid-state equipment it is desirable to have a separate negative and positive voltage supply having a common reference (ground). In this case, a center-tapped transformer may be used in a bridge rectifier circuit as shown in Fig. 8-7. For low ripple content, the filter capacitors run up to 3000 mfd or more. Another fullwave rectifier circuit

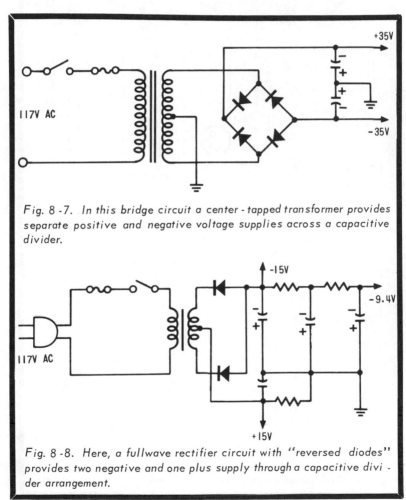

Fig. 8 -7. In this bridge circuit a center - tapped transformer provides separate positive and negative voltage supplies across a capacitive divider.

Fig. 8 -8. Here, a fullwave rectifier circuit with "reversed diodes" provides two negative and one plus supply through a capacitive divi - der arrangement.

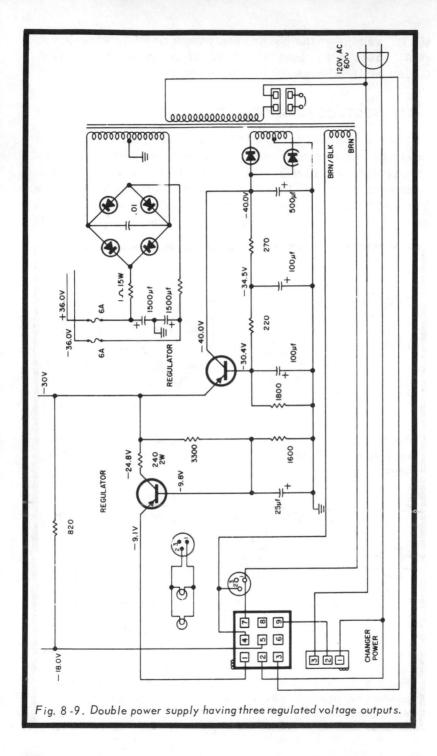

Fig. 8-9. Double power supply having three regulated voltage outputs.

(Fig. 8-8) uses a center-tapped transformer and two "reversed" silicon diodes in a split-voltage arrangement. The center point of a capacitive voltage divider is connected to chassis to provide a "common" reference for separate negative and positive voltage supplies. A large combination AM/FM stereo and phono console may have a double power supply like that shown in Fig. 8-9. The negative-going supply at the bottom has a number of stabilized DC outputs provided through two power transistor regulators.

The better grade table-top AC-operated transistorized radios today, those having separate AM/FM RF and converter stages, will usually have a fullwave rectifier power supply. The less expensive units will have a halfwave rectifier. Some electron-tube type tape recorders are still around. A typical power supply for these is shown in Fig. 8-10. Notice that the power transformer windings are integral with the motor windings. Also notice the "center-adjusted" hum balance control across the AC tube heater supply. These controls can be carefully adjusted while observing a scope screen to eliminate 60-Hz hum.

VOLTAGE - REGULATING ELEMENTS

It is well beyond the scope of this Chapter to give a detailed description and fully characterize the variety of semiconductors, including the zener types, which may be used as power-

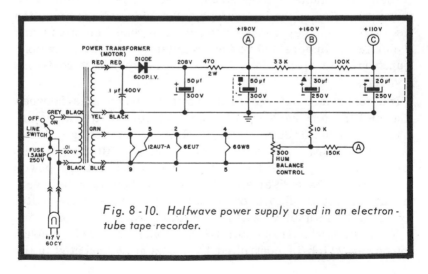

Fig. 8-10. Halfwave power supply used in an electron-tube tape recorder.

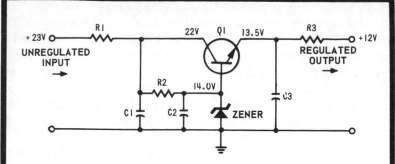

Fig. 8-11. Series-voltage or emitter-follower regulator circuit using a zener diode to clamp the transistor base at 14V (0.5V bias).

supply regulators. We will take a glance only at the regular zener here.

The conventional zener is a specially constructed diode having an extremely high back-resistance up to its critical breakdown point, or zener-voltage limit. (Do not confuse this with the avalanche-breakdown phenomenon—they are two different mechanisms.) When the zener point is reached, the diode's reverse resistance drops sharply to a small value. Current flow then increases, but the voltage drop across the zener remains relatively constant. When the zener is reverse-biased, this characteristic makes it possible for the diode to handle extreme current swings and still provide an essentially stable voltage drop. Each zener manufactured has a specified operating voltage and a wide voltage selection is available. They can be used as reference elements and "clamps," clippers and voltage regulators—alone or in conjunction with transistors and voltage-dependent resistors (VDRs).

If a zener is employed to maintain a steady bias between the emitter and base (or base and emitter) of a transistor, then the transistor can be used to stabilize a low-voltage power-supply's output. The voltage stabilizing and filtering circuit shown in Fig. 8-11 is frequently seen in low-voltage power supplies. It is a "series voltage," or "emitter-follower" regulator. It is also used to perform a "capacitive-multiplier" action which we will not go into here.

To begin with, transistor Q1 dissipates a steady amount of power which is controlled by the zener connected in the base circuit. The transistor, together with R1, R2, and the zener, drop the supply potential from 22 to 13.5v. The breakdown potential of this particular zener is 14v. While the diode conducts at this voltage, C1 and C2 maintain the voltage. Having a base-to-emitter bias of 0.5v, the NPN silicon transistor conducts and provides a regulated potential of 13.5v at the power-supply output. R3 drops the potential to 12v. The transistor also "filters" out a portion of the

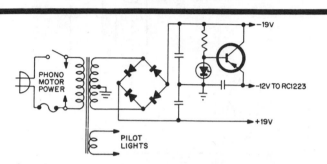

Fig. 8-12 This bridge rectifier circuit provides a separate 19V positive and negative supply plus a regulated and filtered -12V supply. The zener clamps the transistor's base voltage.

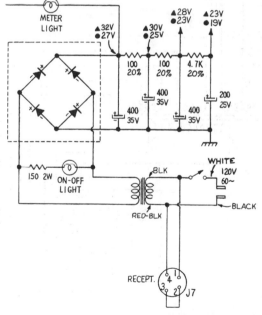

Fig. 8-13. This tape recorder power supply is heavily filtered.

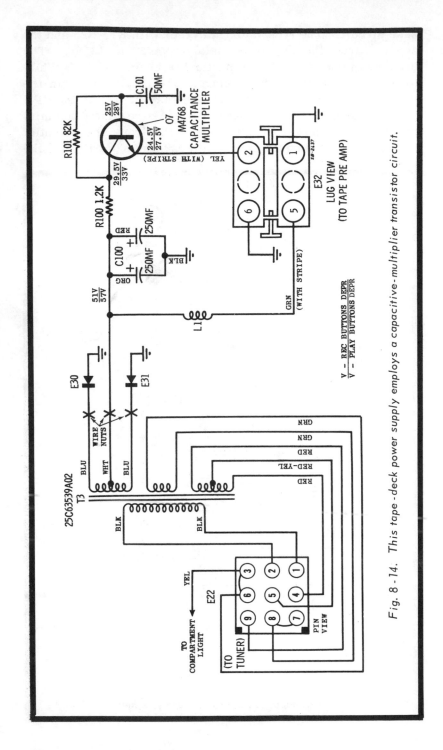

Fig. 8-14. This tape-deck power supply employs a capacitive-multiplier transistor circuit.

ripple. It should be noted that the zener will continue to conduct at 14v even if somewhat less-than-normal voltage appears at the regulator's input. In this case, less voltage is dropped by R1 and R2.

In some respects, a VDR serves somewhat like a zener.

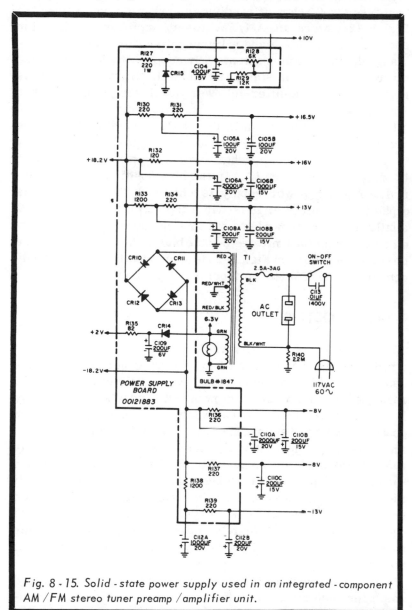

Fig. 8-15. Solid-state power supply used in an integrated-component AM/FM stereo tuner preamp/amplifier unit.

Relatively large currents through a VDR result in only small voltage variations across it. Like the zener, the VDR tolerates current changes through it without much change in voltage.

Let's take a look at a few more power supply circuits. The "split-voltage" capacitive-divider arrangement shown in Fig. 8-12 powers an AM/FM tuner/amplifier combination and has a 19v positive and a 19v negative supply plus a zener-clamped, low-ripple content -12v supply which powers the tuner section. This stabilized and well-filtered -12 volts is provided by an emitter-follower (series regulator) power transistor.

The power supply shown in Fig. 8-13 is employed in a tape recorder. Notice that the measured voltages vary from play to record functions, somewhat lower when the function switch is on record than when it is in the play position. This is normal. Another power supply is shown in Fig. 8-14. This capacitance-multiplier is used in a tape deck and is somewhat similar to others shown here but this one does not use a zener diode to clamp the emitter-to-base bias of the transistor. The power supply shown in Fig. 8-15 is employed in an integrated-component AM/FM stereo tuner preamp/amplifier unit.

TROUBLESHOOTING

Yesterday's power supply was only a means of "powering" a radio receiver or an amplifier. It had generally poor regulation and a high ripple content. Very little equipment used regulated supplies with VR glow tubes. Today's power supply is more complicated and has more components. Although yesterday's power supply always rated high on the check-list when the equipment went dead, today's defective power supply can cause all kinds of fringe-type problems—especially those supplies that power FM stereo tuners, tape recorders, Hi Fi-type preamps and amplifiers. Contrary to the detailed predictions of yesterday's "crystal-ballers" that today's equipment would be infinitely more simple, it has become more complicated. And power supplies are no exception to the rule.

Many power supplies in high-class, medium to high-powered amplifiers use circuit breakers. Power lines also are sometimes fused. Some amplifiers are fused additionally.

As in the older, electron-tube power supplies, what frequently appears to be power supply problems are deep-seated faults in the equipment. For example, if an amplifier is running at a very low power level and the circuit-breaker kicks out, suspect a defective transistor in the output amplifier. Of course, a shorted power supply electrolytic filter or rectifier diode can cause the same problem.

When zeners and transistors are replaced in power supplies we must be careful to follow manufacturers' mounting instructions carefully. Your best bet on most power supply problems is to check the output voltages and compare them with manufacturers' service data. They should not vary more than 5%.

Although hum from power supplies is not the problem it was in electron-tube equipment, it can still happen, usually because of a defective filter capacitor or power transistor in the supply circuit. Use your scope to isolate the hum source. Noise in a preamp can originate also in defective electrolytics and zener voltage-clamping diodes.

CHAPTER 9

Tape Recorders and Players

Tape recorders and players are made in a variety of types and combinations. Tape recorders, for example, come in open reel-to-reel, cartridge reel-to-reel, and continuous-loop cartridge types for general use. Some are portable and operate from batteries. Today's commercial grade tape recorders are generally confined to "open" reel-to-reel types, although at least one continuous-loop cartridge is designed for professional use.

"Straight" tape players come in open reel-to-reel, reel-to-reel cartridge, and continuous-loop cartridge designs for home use; types for either mobile (auto, boat, plane) or home use; combination AM or FM radio/tape arrangements; rechargeable-battery portables (which are somewhat more practical than battery-operated tape recorders) and complete table-top or bookshelf-style integrated-component combinations having tape cartridge players, AM/FM monophonic and FM stereo radio and stereo amplifiers. Some players have even been designed to be used as separate units which operate through existing mobile AM radios. (See Chapter 10 for a discussion on mobile-type players, together with auto radios.)

Almost all general use tape recorder systems made today are either hybrid or solid-state. The hybrids, of course, are obsolescent. Virtually all modern tape-cartridge and cassette players now made have solid-state circuitry. Typical home-type tape players are usually designed to operate as "head ends" for existing Hi Fi stereo equipment, normally being plugged into the tape input of the Hi Fi preamp or integrated amplifier. The mobile and portable types are complete with their own output amplifiers and speakers. Both tape recorders and tape players are electromechanical devices. This Chapter covers "typical" mechanical and electronic operating principles, trouble symptoms, essential adjustments, general trouble-shooting techniques and procedures.

A "TYPICAL" TAPE RECORDER

Because of the wide variety of equipment designs involved, it is impossible to describe as "typical" any existing reel-to-reel tape recorder. And our remarks here are referenced to both tape recorders and tape "decks." A tape recorder is complete in one enclosure, including audio output and speaker system, while a tape deck doesn't have an audio-output amplifier or speaker system. We will also think in terms of two and three magnetic heads, although our discussion is generally confined to two heads—an erase head and a combination record and playback (R/P) head.

As far as service is concerned, all tape recording units can be divided into two sections—the tape transport, or mechanical section, and the electronic section. The electronic section can be subdivided into: (1) the power supply (2) the pre-amp and amplifier (including a speaker or speakers), and (3) the bias/erase oscillator. But before we get into the electronics, let's consider some basic mechanical aspects.

TAPE TRANSPORT

Although most recorders you will be called upon to adjust or "keep alive" will have a single motor, some have two torque-type induction motors—one for each reel shaft. And you may be honored occasionally by a tape recorder which has three motors. But don't let the third motor scare you.

In the two-motored job the take-up reel motor speed is governed by the tape speed across the capstan. The supply reel motor has reverse torque which is inversely proportional to the take-up reel motor. As we can see, this is a motor-drag drive system. In another case, you'll find a spring-belt drive which is connected to both reels. The arrangement provides slippage through spring-stretch and pulley design, which maintains proper take-up and supply action.

A third system, sometimes called the friction-clutch drive arrangement, uses a take-up pulley, a fiber or leather clutch

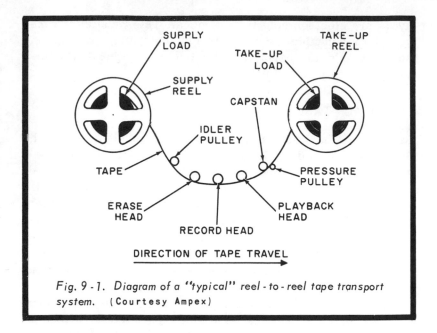

Fig. 9-1. Diagram of a "typical" reel-to-reel tape transport system. (Courtesy Ampex)

disc and felt, a spring, and a tension adjustment screw. The essential drag is provided by the clutch disc. Other systems, including variations of these, may also be employed.

The tape transport of a reel-to-reel tape recorder may have an arrangement somewhat similar to that shown in Fig. 9-1. As indicated by supply and take-up reel placements, the tape travels from left to right. Also notice the R/P and erase head positions. Observe the idler and pressure pulleys (or rollers). When a tape recorder is switched on, the pressure roller fits snugly against the capstan. To illustrate a tape transport design variation, look at the arrangement shown in Fig. 9-2.

From the maintenance viewpoint, we are concerned most frequently with mechanical component wear. This applies to all components subject to friction, contamination, and repeated use—including heads, pressure pads, brake and drive mechanisms, rollers, and function switches.

ERASE AND R/P HEADS

Essentially electromagnetic, but subject to constant mechanical friction, R/P and erase heads are an outstanding trouble spot for maintenance technicians. Although separate record and playback heads are employed on home and general type recorders, they are most often used on commercial grade equipment. But we will be concerned here primarily with combination R/P heads. R/P heads and erase heads may be designed and constructed to mount as separate units or they may be placed together in one shell. We will come back to R/P and erase heads in the maintenance section of this Chapter.

BIAS / ERASE OSCILLATOR

In the tape recorder's electronic section, we are frequently concerned with the bias/erase oscillator and its proper function. First, let's dwell briefly on the need for bias/erase oscillator voltage. As every well-informed technician already knows, hysteresis effects develop when a magnetizing force is applied to a magnetic material—in this case, the oxide deposited on the active side of the recording tape—and this problem naturally calls for some corrective measures. Up until now, the engineers have devised a more or less standard approach to the problem by adding a so-called "supersonic" voltage to the regular audio voltage normally applied to the R/P head during the recording process, just another special-type "countermeasure" used against audio distortion.

This AC bias voltage normally runs about 10 times the applied audio voltage. An erase/bias oscillator's frequency may range from about 30 to 100 kHz. Some will be higher. A larger voltage from the bias/erase oscillator usually drives the erase head to eliminate previous tape recordings, in addition to the bias supplied to the recording head to compensate

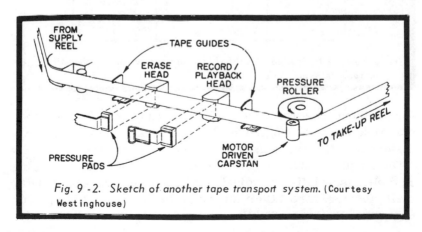

Fig. 9-2. Sketch of another tape transport system. (Courtesy Westinghouse)

141

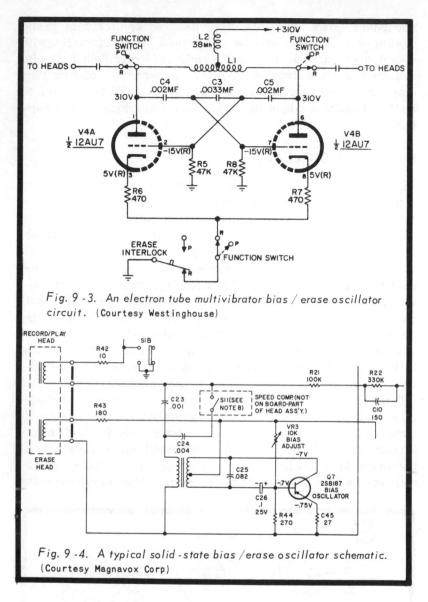

Fig. 9 -3. An electron tube multivibrator bias / erase oscillator circuit. (Courtesy Westinghouse)

Fig. 9 -4. A typical solid -state bias /erase oscillator schematic. (Courtesy Magnavox Corp)

for nonlinear tape response. If no recording bias is used, the playback would be bugged by harmonic and intermodulation distortion. No bias is used on playback, of course.

The electron-tube bias/erase oscillator schematic shown in Fig. 9-3 is relatively typical and is easily recognized as a multivibrator. The LC circuit consisting of L1, C3, C4, and C5, resonates near 60 kHz. But many tape recorders made today are hybrid or completely solid-state. The bias/erase oscillator schematic shown in Fig. 9-4 can also be considered relatively typical. This circuit operates at 35 kHz. Another solid-state bias/erase oscillator circuit is shown in Fig. 9-5. This "push-pull" arrangement operates at 80 kHz. We are not too much concerned here with other electronic circuitry since preamp and amplifier essentials have already been covered in other chapters.

GENERAL CIRCUIT VARIATIONS

Despite the "relatively typical" expression, many monophonic and stereo tape recorder amplifiers do not comply with the "typical" image. For example, while they may be essentially similar, not all right- and left-channel amplifiers can be called duplicates. We remember seeing one tape recorder which had a left-channel amplifier employing its output transistor in the dual role of power output amplifier and bias/erase oscillator. Additionally, the left-channel amplifier of this unit had one more transistor than the right channel. And this extra transistor served as amplifier for a level-indicator meter. So much for the vagaries of the old cliche "typical."

In addition to supersonic bias voltages, many tape recorders

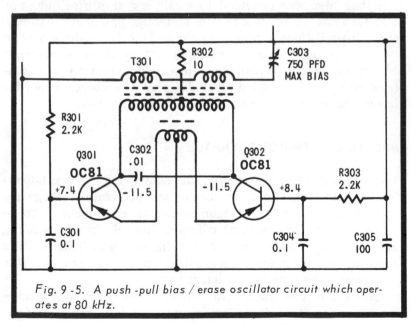

Fig. 9-5. A push-pull bias/erase oscillator circuit which operates at 80 kHz.

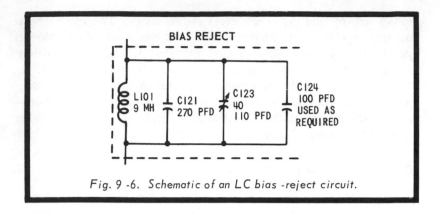

Fig. 9-6. Schematic of an LC bias-reject circuit.

employ negative feedback from the second amplifier stage back to the input stage to provide a better match between the R/P head in both play and record functions. And various feedback levels may sometimes be used for different tape speeds.

Some tape recorder amplifier circuits use tuned traps or "bias-reject" LC circuits to attenuate the bias voltage so it will not interfere with the audio signal. A circuit of this type is shown in Fig. 9-6. In some inexpensive recorders DC bias is used on erase heads, but DC bias is critical and does not provide the best results. In some arrangements, "mute" switches also are provided to "kill" the amplifier output on fast forward or rewind to eliminate the "monkey chatter." Other circuits have switches to remove the erase-head bias so "sound-on-sound" can be recorded. Many other "user-type" conveniences may be provided, depending primarily on equipment cost. A simplified block diagram of a stereo recorder having three separate heads is shown in Fig. 9-7.

MECHANICAL TROUBLE SYMPTOMS

As you will discover shortly, the above heading is misleading. But it is impossible to sharply and accurately categorize trouble symptoms in electromechanical equipment. To be more precise, mechanical defects, like electronic defects, always leave their symptomatic fingerprints behind. But many mechanical defects can cause electronic symptoms. This may sound like a paradox, but it isn't.

Reflect for a moment: An old or misaligned R/P head—mechanically imperfect—can cause poor high-frequency response and low-level output from the recorder amplifier: electronic symptoms. And if the bias/erase oscillator is providing a higher-than-normal bias—an electronic fault—we can be confronted with the same trouble symptoms: either poor HF response, low output, or both. There's nothing new about this, but it does require special emphasis as we approach tape recorder troubleshooting and maintenance.

Suppose a customer brings in or asks you to pick up a tape recorder which suffers from what may be called "wow" and "flutter." In this case, we can come close to 99.9% accuracy by pointing a finger at some mechanical component. But the situation can appear a little complicated because at least a dozen or more mechanical imperfections can cause wow and flutter.

In our discussion of wow and flutter, we may as well go whole hog by adding "drift," like the hair-splitting purists among us. They have described "drift" as a condition which exists when the input signal (tape speed) varies at a rate below 1.0 Hz. Flutter, they say, exists when the input signal varies at a rate of 10 Hz or more. And wow...wow (some purists tell us) occurs when the input signal varies between 1.0 and 10 Hz, but others say "about 0.5 kHz." Most intelligent audio engineers have long ago called these effects "flutter"—including all tape variations from 1.0 to 10 Hz. For practical purposes, then, we can ignore the semantic overtones here and

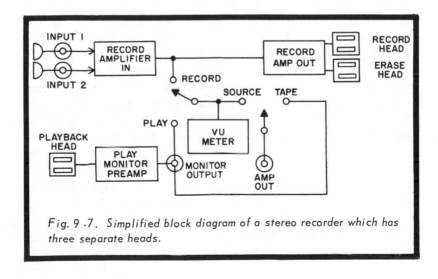

Fig. 9-7. Simplified block diagram of a stereo recorder which has three separate heads.

label all three conditions flutter since all three effects have the same cause.

Drift, flutter, and wow then are simply wavering sounds in the recorded material caused by irregular tape speed and can be detected easily by listening to a test recording which has a steady tone. If our ears are bad and our eyes are only half bad, then we can detect the flutter effect on a scope screen. As previously mentioned, flutter is almost always caused by one or more mechanical faults. These include the effects of dirty tape, drive belt irregularities, changes and variation in AC line voltage and frequency, variations in bearing friction, faulty gearing, worn "rubber" wheels, greasy pressure rollers, too much drag in brakes, a warped tape reel that rubs at some point, misadjusted idler pulley, a binding pressure roller, incorrect pinch between pressure roller and capstan. A few other seldom observed causes may also crop up occasionally.

After these predominantly mechanical causes which frequently produce "electronic" symptoms, we are abruptly projected into other prevalent mechanical failure areas, such as function switches. Whatever the switch type involved, they do wear and corrode under certain circumstances, and they do fail to perform properly. A customer may complain, for example, that his recorder "plays but does not record or erase." Two causes generally create this symptom—a bad bias oscillator component or a faulty play/record switch. The same recorder may play back and record properly, but fail to erase. And in that recorder group having "superimpose" (sound-on-sound) switches, erase can fail because this switch is defective. But the erase head can be defective also. In the combination tape recorder group, a defective microphone/radio/phono switch or a defective play/record switch can cause a unit to play back and erase correctly but fail to record.

In still another case, a customer may push a motor switch "button" on his recorder and hear noise from the speaker but nothing moves in the tape transport system. Here again, the motor switch can be defective or, depending on design, improperly adjusted. But the motor can be defective or an "idler wheel" is failing to engage the motor pulley.

GENERAL MECHANICAL MAINTENANCE

Like most electromechanical equipment servicing, preventive maintenance is the only intelligent approach to tape recorder maintenance. If a unit has been operated for a total of 15 or 20 hours, it needs attention—no matter how good it may appear to be operating at the moment. But since the general public is not usually aware of this need, you must sell the idea. At least, when a unit comes in for repair of whatever kind, you should call the owner's attention to the need for a complete cleaning and check.

If the user is to obtain best results, tape heads, pressure rollers, pressure and drag pads, tape guides and capstans must be cleaned periodically to remove an accumulation of oxide and base material which wears from the tape. This accumulation is frequently mixed with oil, lint, and ordinary dust, preventing the tape making intimate contact with R/P and erase-head pole faces.

Tape heads also need adjusting occasionally. And they will have to be replaced after a modest amount of wear—that is, if the owner wants consistent results.

The previously mentioned parts are cleaned with a "Q-Tip" moistened in alcohol and squeezed out, after which the parts are wiped with a clean, lint-free cloth. Special cleaning tapes, made from absorbent material impregnated with a cleaning substance, also are available. It should be remembered, however, that some manufacturers may warn against using alcohol or other cleaners on their tape recorder equipment. In this case, you should use the cleaner recommended in that manufacturer's service manual.

A few other cautions should be sounded at this point: Do not use carbon-tetrachloride (carbontet) as a cleaner. There are two good reasons for this: (1) Its fumes are highly toxic if breathed. (2) It leaves an undesirable, oxidizing residual deposit on all surfaces. Some people (who should know better), still recommend carbontet as a cleaner. Although they are sometimes recommended (also by those who should know better), do not use pipe cleaners or metal-ferruled brushes to clean tape heads. These can damage delicate head surfaces. Do not use alcohol or any other solvent to clean drive pulleys or belts. Wipe these surfaces clean with a lint-free cloth. If metal surfaces are badly "gummed-up," use a detergent and warm water. Avoid getting alcohol on plastic housings.

When all parts are clean, check the pressure and "drag"

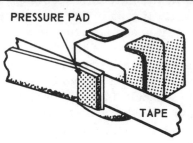

Fig. 9 -8. Pressure pads are used to keep the tape in firm contact with the head pole pieces.

pads. If any pressure pad movement occurs when the tape transport is first started, pads may be loose on their carrier mounts. They may need to be properly positioned and re-cemented or re-glued in place. Pressure pads, made from felt or some type of fabric, press against the back (shiny side) of the tape, forcing it into intimate contact with the head (see Fig. 9-8). Both pressure and drag pads also become hardened and lose resiliency. These should be replaced. If you're ever in doubt about a maintenance procedure, check the manufacturer's service instructions carefully. This applies especially to lubrication since lubricants in the wrong places (on drive surfaces, motor pulleys, belts, flywheels, etc.) can cause a lot of unnecessary grief.

Since all equipment is properly lubricated before it leaves the factory, tape recorders require very little lubrication and only after considerable use. The lubricants specifically recommended by the manufacturer should always be employed. Oil is normally used only on bearings—like clutch, flywheel, turntable, etc. Heavier lubricants are generally recommended for pushbutton latching bars, brake arms, and various levers. One or more still heavier lubricants—depending on the arm or lever design—may be recommended. Once again, refer to manufacturer's specific lubricant recommendations for a particular tape recorder.

In the event a tape recorder requires a new R/P or erase head, consult the appropriate detailed service instructions and parts list for the specific model. And this includes head alignment, adjusting torque screws, pressure roller alignment, and tension adjustments. To measure torque and spring tensions you'll need special inch-ounce scales or a standard torque gauge. Most expert tape recorder technicians use standard test tapes for checking head height and azimuth (tilt) adjustments. Perhaps it will prove helpful at this point to look a little closer at certain important maintenance aspects of the R/P and erase head subject.

HEAD WEAR, CARE, AND ADJUSTMENT

First, a few words on the question of head life. Disregarding head design factors relating to the "toughness" of a head, perhaps the most important factor is whether or not the tape transport is designed to use pressure pads. Generally speaking, head life can be as low as 400 hours or less when a certain type of pressure pad is used (one having small area contact at a pressure of more than a half-ounce). On the other hand, the same head may last upward of 4000 hours if the equipment is designed to operate without pressure pads. Of course, many other factors are involved in determining head life. For example: (1) Quality, type, or condition of tape used. (2) Speed of tape (ips) across the head face. (3) How often the head and other tape contact parts are cleaned. (4) Condition of pressure pads and kind of material they are made of. Other factors also may be involved.

When the azimuth alignment of a tape head is off—when the head gap is not perpendicular to the tape travel—a serious loss of high-frequency response may occur. Also, the tape will have a tendency to "skew" upward when the head is tilted backward, as shown in Fig. 9-9. And even if the azimuth alignment is perfect, similar intermittent trouble can develop from warped or wobbly reels or worn tape guides. Vertical tape walk, or "skewing," as shown in Fig. 9-10, can cause the same effect on frequency response as a tilted head. And azimuth alignment is always necessary after a head is replaced.

To emphasize what can happen to frequency response either on record or on playback when close contact is not maintained between tape and head, observe the graph in Fig. 9-11. Notice how much the higher frequencies are attenuated when the spacing between head and tape increases only a very small amount.

Since we can reasonably say that the care and the life of tape recorder heads are interdependent—other factors considered normal—we must first keep all tape contacting components clean (as previously described), in addition to cleaning the heads after every 10 or 15 hours of use. And, as previously

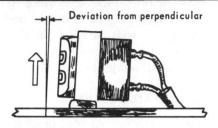

Deviation from perpendicular

Fig. 9 -9. *When the head gap is not perpendicular to the tape, a serious high-frequency loss results.*

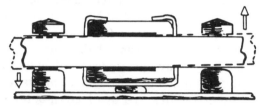

Fig. 9 -10. *Excessive reel wobble and worn tape guides can cause excessive skew in directions indicated by arrows.*

made clear, other factors can make this need for cleaning even more frequent. For example, new tapes will build up oxide deposits much faster than old tapes which have had a large portion of the oxide polished off after repeated use.

When you clean the heads and tape guides, make sure the head covers are removed according to the manufacturer's instructions. Don't attempt to clean heads and tape guides through the narrow tape threading slot. Incidentally, the head covers should be cleaned on the inside at the same time.

Once the heads are perfectly clean, they should be examined under a suitable magnifying glass to determine the extent of wear. In many cases, as heads wear the core gaps widen. Compare the gap width to that in a similar type new head. Tape guides also should be examined in the same way. Badly worn heads and tape guides should be replaced.

The tools used to repair or adjust tape recorders, particularly those wrenches and screwdrivers employed in making head adjustments or replacing heads, should never be used for any other purpose, especially not on radios and TVs. A magnetized tool can magnetize a record or R/P head.

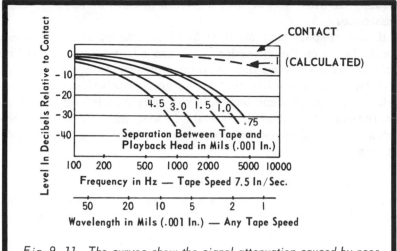

Fig. 9-11. The curves show the signal attenuation caused by poor -R/P contact. Notice the greatest attenuation appears in higher frequency areas.

HEAD AND TAPE GUIDE DEGAUSSING

When a tape recorder is in the record position, the record or R/P head becomes an electromagnet. After a few hours of recording, the head begins to retain a small amount of magnetism. Tape guides and erase heads may also become magnetized. After considerable use in the record position, this residual magnetism builds up to a point where the high-frequency response of the head is reduced. And a magnetized record or R/P head will record noise on the tape—even when the recorder is being used in the play-back position. This makes it necessary to periodically degauss all the heads, metal tape guides, or other metal parts adjacent to the heads, after every cleaning. Degaussing should be done too, after making measurements on heads—especially after coil continuity checks with a VOM. This is a process somewhat similar to that employed in degaussing color picture tubes. Recording head degaussers, of course, are very small compared to those used on color TV tubes. Degaussers are available from the parts supply department of each tape recorder manufacturer, and tape recorder or general electronics distributors. The degaussing instrument tip should be covered with a soft plastic coating to prevent any metal coming in contact with the heads. It also should have a switch in the line cord for convenience.

If you suspect that an R/P head is magnetized, put a new reel of tape on the recorder, turn the record level control up full and without recording anything through the microphone let the tape run for a few feet in the record position. Fast-rewind and play the tape back. If you hear popping or "hissing" on the playback, demagnetize the heads, guides, and all surrounding metal parts. If your ears are not so good, then do it this way:

(1) With the recorder off and no tape in it, connect your AC VTVM across the speaker (or proper load resistor) of one channel and set the meter on whatever scale is necessary for the particular amplifier.

(2) Switch the recorder on, depress the playback button and adjust volume and tone controls full up.

(3) Observe if there is any reading on the VTVM. If so, note how much and proceed to Step 4. If not, move the meter switch to whatever lower scale provides a reading. Note the voltage.

(4) Switch the recorder off, switch the VTVM back to the higher scale and load a new, unused tape in the recorder.

(5) Switch the recorder back on, and with the controls and play-back button in the same position, observe if the VTVM reads any voltage If not, again move to whatever lower scale necessary to provide a reading. Note the voltage. Repeat this process on the other channel.

If the voltage readings noted in Step 5 increase considerably over that in Step 3, on either or both channels, the heads need degaussing.

A number of procedures have been described for properly degaussing heads and all surrounding metal parts. We will describe one method here: (1) Plug the degausser into an AC receptacle. (2) Place the degausser tip on the head face or against tape guides. (3) Switch on the power to the degausser. (4) Move the degausser tip up and down the head or tape guide a few times. (5) Slowly move the degausser straight away from the head to a distance of about 5 feet. (6) Switch off the degausser power.

HEIGHT AND AZMUTH ADJUSTMENTS

The term "height adjustment" should be self-explanatory. The head must be in line with the capstan/pressure-roller/ tape-guide configuration so the recorded tracks on the tape fit properly over the R/P and erase-head gaps. The heads cannot be too high or too low. This is particularly true of 8-track systems where the tape tracks and gaps are narrow.

The azimuth adjustment is a means of making the head point straight up and down—in the vertical plane—at right angles to tape travel, perpendicular to the tape edge. The heads must not tilt in any direction and must not be turned right or left. It should be noted that some erase heads are fixed to their mounts and have no azimuth adjustments. And the combination stereo head shown in Fig. 9-12, designed for 8-track recorder systems, combines both R/P and erase head in one shell to simplify head adjustments on critical 8-track systems.

It is not recommended that you attempt to make these adjustments by physically aligning the heads with the tape. The proper way is with special test tapes which are available from tape recorder and tape head manufacturers (Ampex, RCA, GE, Nortronics, and others). You will also need service instructions to locate the adjustment screws.

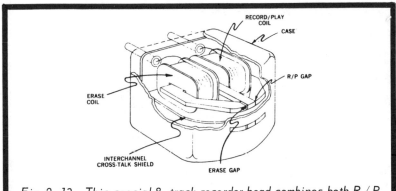

Fig. 9-12. This special 8-track recorder head combines both R/P and erase coils in one shell to simplify head adjustments on critical 8-track systems. (Courtesy Nortronics)

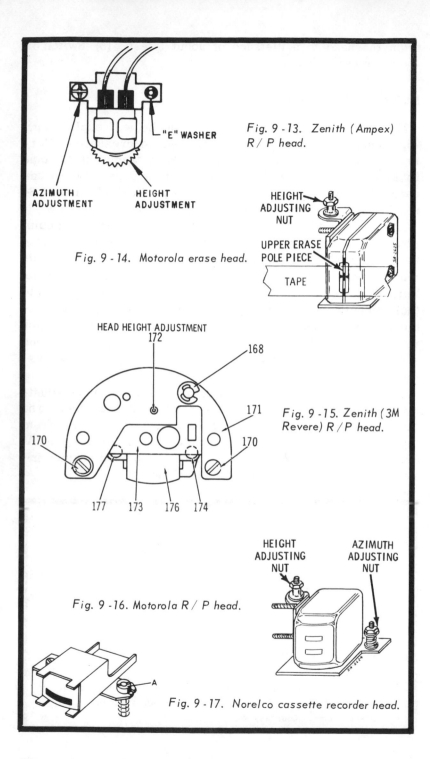

"E" WASHER

Fig. 9-13. Zenith (Ampex) R/P head.

AZIMUTH ADJUSTMENT   HEIGHT ADJUSTMENT

Fig. 9-14. Motorola erase head.

HEIGHT ADJUSTING NUT

UPPER ERASE POLE PIECE

TAPE

HEAD HEIGHT ADJUSTMENT
172

168

171

170   170

Fig. 9-15. Zenith (3M Revere) R/P head.

177   173   176   174

HEIGHT ADJUSTING NUT   AZIMUTH ADJUSTING NUT

Fig. 9-16. Motorola R/P head.

A

Fig. 9-17. Norelco cassette recorder head.

Drawings of a few representative R/P and erase heads, indicating adjustment points, are shown in Figs. 9-13 through 9-17. Always adjust the azimuth <u>first</u> and the height after. Replacement heads are available from tape head manufacturers' distributors throughout the country. When replacing a head, it is recommended that it be replaced complete— with a new mount.

CHECKING RECORD BIAS AND ERASE HEAD CURRENT

The amount of voltage generated by a bias/erase oscillator varies according to a number of factors, the most important is the oscillator's frequency. And the amount of normal current and voltage appearing in or across the record or erase head windings will also vary. "Typical" current in an R/P head is from around 0.3 to 1.3 ma. Erase head current may run from about 12 to 40 ma in different brands and models. To check these voltages and currents, the manufacturer's service instructions should be carefully followed.

For illustrative purposes, let's consider a general procedure for current measurements in an R/P and erase head: First, the R/P head procedure (see Fig. 9-18A): With the recorder switched off, adjust the record level control to minimum. (2) Disconnect the ground lead (or leads) from the R/P head and insert a 10-ohm 5% tolerance carbon resistor in series with the lead. Reconnect the ground lead. (Some tape recorders have these resistors permanently installed.) (3) Connect an accurate audio type AC VTVM across the resistors. The hot (red) meter lead should be next to the head. (4) Switch the recorder on and depress the play/record button. (On stereo R/P heads each half of the head winding must be measured as shown in Fig. 9-18A and both the right- and left-channel record buttons must be depressed.) (5) While the recorder is operating, read the voltage across the resistor or resistors.

A "typical" reading here may be 0.005v (5 mv) across each resistor, or 0.5 ma of current through each half of the coil, (using Ohm's law). Some recorders may measure slightly less. Another solid-state stereo recorder, made by another manufacturer, may read 7 mv, or 0.7 ma in both sections of the R/P head. And some manufacturers may recommend using a 100-ohm instead of a 10-ohm resistor.

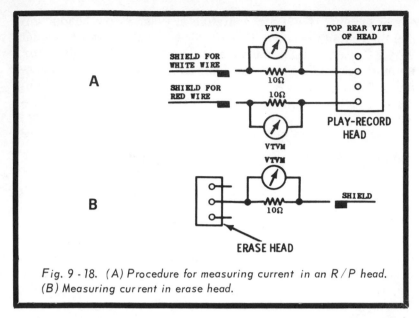

Fig. 9 - 18. (A) Procedure for measuring current in an R / P head.
(B) Measuring current in erase head.

Variable resistors or capacitors (screwdriver-adjustable), are usually connected in series with the bias oscillator output to the record heads in each channel for adjusting the bias properly. If these adjustments do not provide the proper bias, check the bias/erase oscillator circuit for defective components.

The erase head current is measured in approximately the same way (see Fig. 9-18B): (1) Switch the recorder off. (2) Disconnect the ground lead and insert a 10-ohm 5% tolerance resistor as in the R/P head procedure. (3) Connect an AC VTVM as before. (4) Depress the record button and read the voltage as before.

A "typical" reading across the resistor here may be 0.04v, or 40 ma. A plus or minus 4 ma variation margin is permissible. An electron-tube recorder manufacturer recommends using a 100-ohm instead of a 10-ohm resistor. In this case the potential across the resistor reads 2.1v, or 21 ma. It is quite all right if the current is off plus or minus 7 ma.

While making these measurements in solid-state equipment, be careful not to short the erase head connections—the bias/ erase oscillator transistor or transistors may be damaged.

Some electronic adjustments and maintenance checks include record level indicator and bias trap adjustments. And the procedure for checking sensitivity and frequency response of a recorder, as examples, are too lengthy to provide step-by-step details here. These procedures are fully explained in the manufacturer's service instructions or in test instrument application manuals. Likewise, the mechanical adjustments.

CARTRIDGE RECORDERS

A number of cartridge-type stereo tape recorders have been on the market for some years. Their mechanical sections do not differ too much from those used in reel-to-reel types. But the design of different cartridge-types has contributed to the development of dissimilar cartridge transport arrangements. Some cartridge-type recorders operate at two speeds, 1 7/8 and 3 3/4 ips, while others operate at one speed—either 1 7/8 or 3 3/4 ips. We will take a brief look at the RCA Victor cartridge, a portion of one transport mechanism, and the front deck view of one recorder which uses this particular cartridge (basically reel-to-reel enclosed).

An opened view of the clear plastic cartridge is shown in Fig. 9-19. The two sides, A and B (or bottom and top), are held together by two screws when the case is closed. The case contains a spacer, take-up hub and supply hub, braking

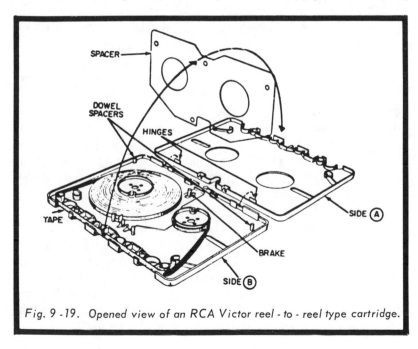

Fig. 9-19. Opened view of an RCA Victor reel-to-reel type cartridge.

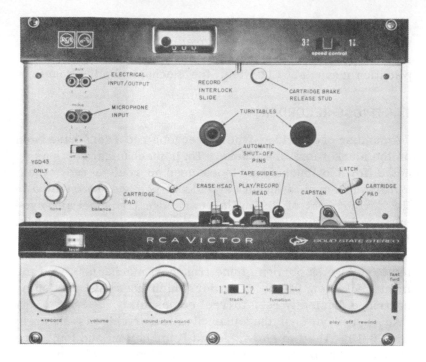

Fig. 9 -20. Front view of an RCA Victor cartridge tape recorder.

mechanism, and two automatic shut-off devices. The cartridge
is available also with interlock actuating barriers so a blank
tape can be recorded or an old tape can be erased and used
for re-recording. These barriers have to be in place before
the record switch can more. Cartridges of prerecorded tapes
have these barriers removed.

Holding up to 600 feet of 1/4" 1-mil tape, the cartridge has
a record/play time (4-track system) of 1 hour on stereo at
3 3/4 ips. On mono this is 2 hours. Playing time can be
doubled at 1 7/8 ips by using the proper tape and recording on
both sides. The cartridge can be turned over and operated on
either side for double recording or playing.

Unlike the "plug-in" type cartridge, this one fits flat against
the transport deck (see Fig. 9-20). Through openings in one
side (or edge) of the plastic case, the tape contacts the capstan, R/P and erase head. The position of the tape can be
viewed at any time and the cartridge can be inserted or removed at any point in the playing time. Brake clamps hold the
tape hubs tight when the cartridge is not being used.

TAPE TRANSPORT

A four-pole, shock-mounted, permanently lubricated motor drives the tape transport. A fan on the motor shaft cools the motor and helps cool the amplifier output transistors. A partial view of the mechanical components of the tape transport when it is in the "play" position is shown in Fig. 9-21A. As

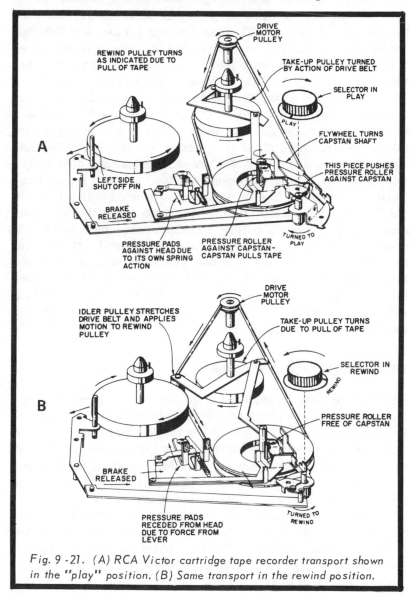

Fig. 9-21. (A) RCA Victor cartridge tape recorder transport shown in the "play" position. (B) Same transport in the rewind position.

in some reel-to-reel tape recorders, the slip-clutch principle is used here to provide the proper speed for the take-up hub as it accumulates more tape. The take-up hub's speed must decrease slightly for each turn of tape it accumulates.

When the mechanism is in the record or play position, as shown, power is transmitted through the belt from the motor pulley to the flywheel and capstan shaft. In this position the belt also drives the take-up pulley in the direction indicated by the curved arrow. The capstan and pressure roller pull the tape across the R/P and erase heads. The capstan, pressure roller, and pressure pads are snug against the tape but the idler pulley (behind the take-up pulley) is not engaged.

Now notice the changes that take place when the recorder is in the rewind position (Fig. 9-21B). Through action initiated by the selector switch, the idler pulley is now engaged with the belt—stretching it out away from the take-up pulley and against the rewind pulley. The pressure roller has also moved away from the capstan, allowing the tape to move freely (forward or reverse). In this position the rewind pulley turns in the direction indicated, pulling the tape from the hub connected to the take-up pulley. The pressure pads are not in contact with the tape and the brake is released. Notice that the motor drive pulley has two steps for both normal and higher speed. The belt is moved from one step to the other through action of the speed control. Refer to manufacturer's service data for more information on this transport and for electronic section details.

VIKING CARTRIDGE RECORDER

What is normally considered to be the original, regular endless- and moebius-loop type cartridge is made by the communications division of TELEX under the Viking brand name. A Viking 8-track cartridge tape recorder/player is shown in Fig. 9-22. TELEX manufactures manual, semiautomatic, and automatic cartridge transport mechanisms and presently supplies transports to a number of auto manufacturers.

Three different Viking tape cartridges are shown in Fig. 9-23. These are available in both standard loop, single-coated 1 mil or double-coated 1 1/2 mil moebius-loop tape arrangements. The largest moebius-loop cartridge handles a maximum of 1200 feet of tape having a playing time of 64 minutes at 7 1/2 ips and 128 minutes at 3 3/4 ips (based on half track monophonic mode). The largest standard loop cartridge carries a maximum tape length of 1500 feet with playing times of 40 and 80 minutes, respectively, for 7 1/2 and 3 3/4 ips in the same half-track monophonic mode.

NORELCO "CASSETTE"

The word "cassette" in French means "cartridge." The Norelco (North American Philips Co.) enclosed reel-to-reel cassette is similar to the RCA Victor reel-to-reel cartridge, except it is smaller and uses 1/8" wide tape instead of 1/4". This makes it possible to design very compact recorders and players, taking up approximately the same space as an average size book. The speed used is 1 7/8 ips. Employing 4 tracks, but unlike the alternate track system of standard 4- and 8-track continuous-loop cartridges (employing tracks 1-3, 2-4; and 1-5, 2-6, 3-7, 4-8, respectively), the cassette uses adjacent tracks 1-2 and 3-4 for stereo.

Both ends of the tape in a cassette enclosed reel-to-reel system are permanently attached to the two hubs, and when the end of the tape is reached the hubs stop turning, either through automatic shut-off or through slip-clutch action which

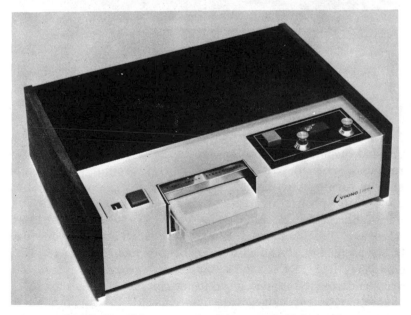

Fig. 9 -22. Viking cartridge tape recorder./player.

Fig. 9 -23. Viking continuous - loop cartridges.

Fig. 9 -24. Norelco cassette recorder /player.

prevents unnecessary pull on the tape. The tape can be fast-rewound, reused, or be recorded or played on the opposite side by turning the cartridge over. Cassettes can be used on either monophonic or stereo Norelco-type recorder/players manufactured by Norelco or by others under license from Norelco. The photo of a Norelco cassette stereo recorder is shown in Fig. 9-24 with microphones and speakers. This company, like TELEX and a few others, provides excellently prepared service manuals for its equipment. A "compatible" cassette-type cartridge made by a cartridge manufacturer is shown in Fig. 9-25.

HOME CARTRIDGE PLAYERS

Although a few vague indicators have prompted some premature predictions from industry "crystall-ballers," it is virtually impossible at this point in time to predict what will happen tomorrow in the cartridge tape player industry. No standardizing trends appear on the immediate horizon. Only one thing can be predicted with any degree of certainty at this moment: We can look forward to continued, standardized uncertainty.

Home-type cartridge players, whether reel-to-reel or continuous loop, come in a number of versions and presently in two basic types. One type is an integrated unit having an amplifier and internal or external speakers. The other type is a "deck" having sufficient output power to drive the amplifier in an existing Hi Fi package. These tape player units may be installed as "drop-ins" under the lids of some existing "theatre-type" or other consoles or be mounted on top of integrated-component or other package-type consoles.

Norelco manufactures an automatic cassette changer (see Fig. 9-26) which functions essentially like an automatic record

Fig. 9 -25. 8 -track stereo tape cartridge.

Fig. 9 -26. Norelco automatic stereo cassette changer.

changer. It handles six cassettes which can be, like the RCA Victor cartridge, turned over to double the playing time.

Basic service procedures for tape players are essentially similar to those of open reel-to-reel type tape recorders. A number of manufacturers provide adequate coverage in their service manuals but others are skimpy. Some manufacturers (or suppliers) have unfortunately oriented their service manuals toward equipment owners!

"TYPICAL" 8-TRACK STEREO DECK

Let's look briefly at a "composite" 8-track stereo "deck" which is commonly merchandised (at this writing) under a wide variety of brand names. They are made in integrated units, containing internal or external speakers. The deck we have in mind is made in two models. One is designed for "drop-in" console mounting and the other for top-of-the-set installation. These units are basically similar to many cartridge players employed in automobiles. They operate at a speed of 3 3/4 ips. All are provided with both manual and automatic track-changing mechanisms.

The cartridge used in these decks contains one tape hub, or "platform," a tape guide, pressure pads, and a pressure roller as shown in Fig. 9-27. The tape feeds from the reel

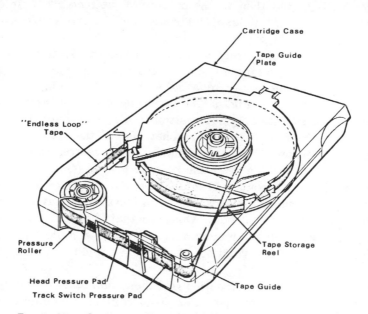

Fig. 9-27. Computron "compatible" cassette-type cartridge.

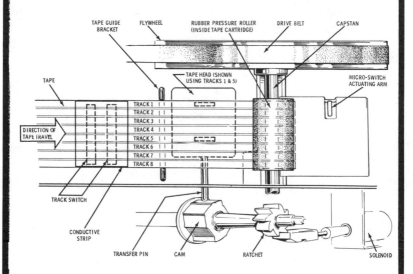

Fig. 9-28. Tape drive and head-positioning mechanism on an 8-track continuous-loop cartridge player. (Courtesy of Motorola)

center past the cartridge tape guide, pressure pads, pressure roller, and then to the outside of the tape reel. In effect, the tape is endless and runs continuously in the same direction.

When the end of a program or set of two tracks is reached, a metal foil on the tape makes contact with the track changing switch—applying power to a solenoid. The spring-loaded solenoid shaft (see Fig. 9-28) is pulled in for an instant and then returns to its original position. A paw on the end of the solenoid shaft detents the ratchet end of the cam shaft to another position. The head transfer pin moves the head up or down. When the last two tracks have played (4-8), the head is moved back to contact tracks 1-5 again. Notice that similar action sequences also take place when the manual program selector button is pushed. Observe that the cam has eight stepping surfaces and moves only 180° for a complete playing cycle. A full 360° rotation of the cam provides two playing cycles.

Complete service instructions for these and other home-type cartridge players are available from various manufacturers, distributors or suppliers, including General Electric, Lear Jet, Motorola, Norelco, RCA Victor, TELEX, Zenith Radio Corp., and others. Also see Chapter 10 for additional information on tape players.

CHAPTER 10

Mobile Radios and Tape Players

Inevitably, if you're in the audio business, you'll be asked to service mobile radios and tape players that may be installed in cars, trucks, boats, or planes. Modern mobile equipment includes "straight AM radios," "straight" FM radios, combination AM/FM radios, FM stereo radios, combination AM radio/stereo tape players, combination FM stereo/stereo tape players and also combination AM/FM stereo/stereo tape players. Additionally, let's not forget the single stereo tape player, some designed for "carry - home" use by merely removing the unit from the vehicle and carrying it home to be plugged into existing Hi Fi amplifying equipment. Finally, there are some older "carry - home" regular AM radios still around. Some radios also use "reverberation" (reverb) units. And some have alerting systems consisting of a transistorized audio oscillator that gives an audible indication through the speaker when parking or headlights have been left on.

For American-made cars, almost all of this equipment is designed for 12-14v negative-ground operation. Some equipment has a polarity-reverse switch or a special plug so it can be installed in foreign-made cars.

GENERAL CONSIDERATIONS

To begin with, radios and tape players, designed for either in - dash or under - dash installation, must be removed from their mounts before most repairs can be made. After repair, the equipment must be re-installed and checked out. On most radios, the important after - repair checkout points include (1) adjusting the antenna trimmer for peak sensitivity (this is normally done at the high-frequency end of the AM band while tuned to either a weak station or between-station noise, but some service instructions specify that the adjustment should be made somewhere between 600 kHz and 1 MHz); (2) checking the antenna by "flipping" it two or three times. This means you grasp the top end of the antenna, flex it back about eight or ten inches and let go. If there's intermittent contact at the antenna base it will usually show up on this check: (3) tune the receiver across the entire band to determine if the stations are uniformly strong. If possible, this check should be done while the owner is present and he should be asked if the radio is working as well as it did before the repair was made.

Some older electron-tube auto radios could often be satisfactorily repaired by replacing a tube (usually an OZ4 rectifier) or the vibrator, without having to remove the radio from its mount (which was, and still is, quite a task on some cars), but few repairs can be made on today's mobile radio or tape player without removing them from their mounts and setting them up on the shop bench. There are some exceptions, however: (1) a defective antenna, (2) a blown fuse, or (3) perhaps a battery problem. And the worst battery problem that you can run into is a "reverse-charged" battery. In this event, the radio won't work and is sometimes damaged. If you ever remove a "dead" radio from its mount and it works perfectly after it's set up on the bench, go check the battery polarity with a meter before you waste time checking the radio. Batteries do become "reverse-charged" at some garages and battery service stations.

Before we get too deeply involved in the pecularities and special problems of mobile radios and tape players, let's understand that successful servicing in this area requires a specialized approach. That is, you need special knowledge. Most of this you can get (after a few years study), from equipment manufacturers' service manuals. You need a good, highly stabilized, well-filtered power supply (a 12v battery will work nicely if you have a heavy-duty battery charger and service the battery periodically); a separate work-bench area where one or more permanently mounted test speakers having convenient clip-leads (or plugs) are located; a few special tools and a proper stock of parts, including special fuses and replacement antennas. The well-planned shop also has an auto "drive-in" area where the equipment can be conveniently removed and reinstalled.

Although many less-expensive AM/FM mobile radios are designed so some stages are shared between the two modes, more expensive types employ separate stages from the RF, through the IFs and up to the first audio stage. The block diagram of an AM/FM auto radio having separate stages for both operating modes is shown in Fig. 10-1. We should remember also, as mentioned in Chapter 3, that most AM mobile radios have an IF frequency of 262 kHz (or 262.5 kHz) instead of 455 kHz.

SERVICE LITERATURE

The need for detailed manufacturers' service instructions on mobile radios and tape players is even more important than it is for strictly home-type equipment. This is particularly true for service details on equipment removal, reinstallation, and detailed instructions on new equipment installations. Not all new cars and trucks sold have original radio or tape player equipment. Proper schematics and other information is also necessary because a number of essentially similar type radios, having some important variations—including different bench-test hookups—are especially adapted for use in different type cars and trucks. A manufacturer's bench setup instructions for a certain model radio, especially adapted for one car, is shown in Fig. 10-2. Some manufacturers also furnish detailed removal instructions. An example of these directions is given here and shown in Fig. 10-3.

A standard instruction warns: Remove battery ground cable before attempting to remove the radio. Then (1) remove (A) two front, upper phillips screws and three lower 1/4" screws from the ash tray and set it aside. Remove two phillips screws from the center AC (air conditioner) duct and lift the attached air plenum out. (2) Remove (B) two phillips screws from the bottom of the radio mounting bracket, two 3/8" nuts from the side of the radio mounting bracket and remove the bracket. (3) Disconnect (C) antenna and speaker connectors, stereo and rear speaker connectors. (4) Remove (D) knobs, washers, hex nuts, and escutcheon knobs from the radio control bushings. Drop the radio down and remove through the ash-tray opening. After you remove, repair, re-install (or install) a few dozen, you begin to bypass some detailed in-

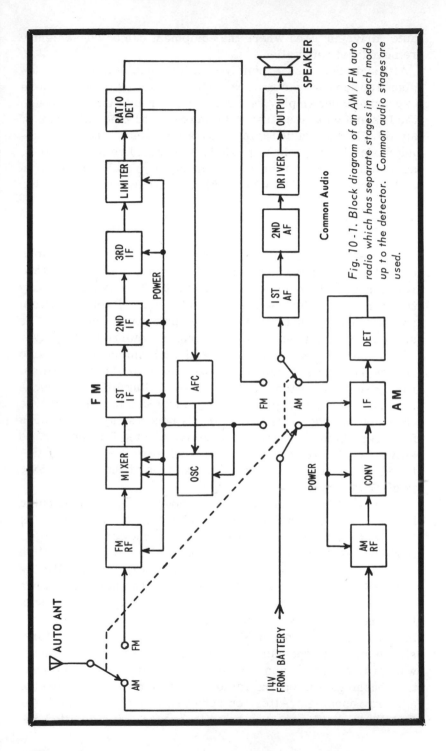

Fig. 10-1. Block diagram of an AM/FM auto radio which has separate stages in each mode up to the detector. Common audio stages are used.

structions as you recognize the standardization trends inherent in the various manufacturers' products.

SERVICING MOBILE RADIOS

In addition to previously mentioned service considerations peculiar to mobile radios, there are a few other specialized problems. First, recently made mobile radios have gone "mod." Complete circuits are encapsulated in modules. Key test points make it relatively easy to isolate trouble to a particular module. This, of course, alters the conventional troubleshooting and repair techniques explained in some other Chapters of this book.

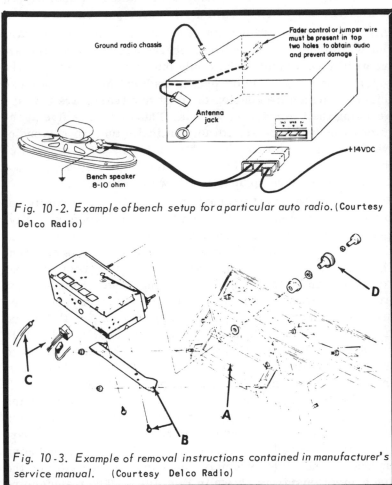

Fig. 10-2. Example of bench setup for a particular auto radio. (Courtesy Delco Radio)

Fig. 10-3. Example of removal instructions contained in manufacturer's service manual. (Courtesy Delco Radio)

Pushbutton and automatic tuning assemblies and circuits usually provide the two prime problems in present day and older mobile radios. Since some improvements have been made in these two systems from time to time, many variations exist in different models which make exact - model service manuals desirable. But basic concepts remain unaltered.

In considering pushbutton tuners, first recall that mobile radios generally employ permeability-tuned circuits and manual-tuning shafts which may have different coupling methods, including one which uses a worm-gear/anti-backlash drive system. Dial - pointer cords may or may not be employed. These tuning methods provide a much more rugged and stable tuning system than used in home-type radios. Generally, a mobile radio may have from four to seven pushbuttons.

Although mobile-radio pushbutton tuning systems may vary somewhat, most of their basic characteristics are similar. But we'll take a brief look at only one specific example, the system employed in Delco mobile radios. Millions of these radios, manufactured during the past few years, are now approaching their service-need peak. This radio has five pushbuttons which can be set to automatically tune in any one of five AM radio stations. Each button has a "slide assembly" which contains a locking lever, a cam, a flat spring, and a coiled button-return spring, as shown in Fig. 10-4.

Any station that is first tuned in manually can be easily set up for pushbutton selection. When the dial is turned manually, the treadle (see Fig. 10-4) rotates to a certain position for the station desired. The clutch assembly holds it in this position. When the pushbutton is pulled all the way out (about 1/2" out from its normal position), the locking lever releases the cam. Now, when the button is pushed all the way in, three functional effects occur: (1) The cam moves parallel to and against the treadle; (2) the locking lever is actuated to hold the cam in a fixed position and (3) the pushbutton returns to its normal position when the thumb is removed. Thereafter, whenever this particular button is pushed all the way in, it actuates the treadle and in turn the tuning-core bar, which causes the same station to be tuned in. It should be noted that declutching action takes place when a button is pushed part way in and released. The main parts of this tuner, including pushbutton and slide assembly, and an exploded view of the clutch mechanism, is shown in Fig. 10-5. The manufacturer's service manual provides detailed instructions on pushbutton and slide assembly replacement, dial pointer calibration and replacement, tuning coil replacement, and clutch adjustments.

Certain models of this radio contain the "Wonder Bar" automatic tuning (signal-seeker) system. Although some earlier models (both tube and transistor) employ a 12AL8 tube in the automatic tuning "trigger" circuit (some circuits having three or four diodes additionally), later models use three transistors, three "regular" diodes, and a zener diode in the automatic tuning circuit. Depending on the model year, this circuit contains a variety of transistor types. To properly service this circuit—to cover all variations that have taken place over a period of some years—service manuals are required for each model year from about 1963 to date.

A brief review of the basic operation of this automatic tuning circuit may prove helpful. Refer to the 3-transistor schematic shown in Fig. 10-6. It should be understood, however, that many small variations exist in these circuits, including those changes made during the past two years. This radio is provided with both a bar-type start switch and a foot switch. Unlike some previous models, both switches provide instant muting. D1 is a muting isolation diode. D2 is for AGC isolation in the sensitivity circuit.

The primary circuits of the automatic tuning section—es-

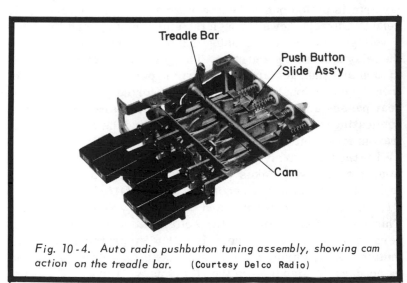

Treadle Bar

Push Button
Slide Ass'y

Cam

Fig. 10-4. Auto radio pushbutton tuning assembly, showing cam action on the treadle bar. (Courtesy Delco Radio)

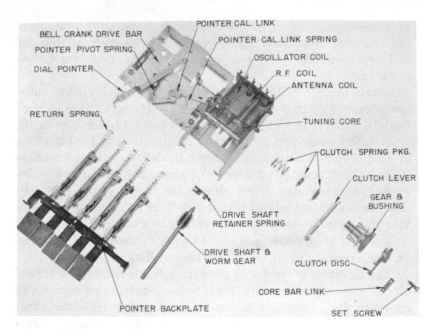

Fig. 10 -5. Main parts of pushbutton tuner with exploded view of clutch mechanism. (Courtesy Delco Radio)

sentially electromechanical—include starting/hold, stopping, audio muting, and sensitivity. When either switch is closed, an electrical circuit is completed through the relay. The relay arm is pulled out of the governor and 14v is applied to the relay. Current flows through the emitter of Q2 and produces a voltage across R1, causing Q3 to conduct. Q3 maintains the relay in an energized state. The governor gear train starts to turn and the rack and worm gear is pulled toward the tuner front by a previously stretched power spring. When the worm gear passes across the flat anti-backlash gears, they turn— duplicating the effect of manual tuning—and the station pointer travels from left to right across the station indicator dial.

When the tuner arrives at the exact frequency of a strong station, the following takes place: A relatively strong signal from the IF transformer secondary feeds to Q1's base, through C1, causing Q1 to conduct and develop a voltage across R3. This cuts off Q2, causing the voltage across R1 to drop to zero, cutting off Q3. The relay arm stops the tuner. Simultaneously, during the tuning process and before Q1 begins to conduct, a signal from the IF primary, through C2, goes to the base and emitter of Q1. The voltage to the emitter is rectified by the AGC trigger diode, D3, and acts as a "bucking," or AGC voltage. The effect of this, together with the IF primary signal voltage on the base, is to keep Q1 shut off until the exact frequency of the station is reached by the tuner. Otherwise, the tuner would stop before a strong station is properly tuned in.

When the worm and rack reaches the front of the set, the "turn-on" tab actuates the re-cycle solenoid switch, energizing the solenoid which now returns the worm and rack to the rear of the tuner again, stretching the power spring in the process, and the dial pointer is pulled back to the low-frequency end of the dial by the treadle - return spring. The treadle bar arm strikes the turn - off tab, the solenoid is switched off and the tuner is inoperative until the start switch is again closed.

Admittedly, the aforementioned is an oversimplified explanation of the electromechanical functions of the receiver, but it is doubtful if a more detailed version (probably available from the manufacturer), would help our batting average. The important thing is how to locate trouble in the electronic and mechanical sections of the tuner. In some trigger circuits the transistors are all directly coupled. In others, Q1's collector is coupled to Q2's base through a 2-mfd electrolytic capacitor. The zener diode clamps the 8v line for stable voltages to the transistors.

TROUBLESHOOTING AND PRECAUTIONS

The Delco radio under discussion here operates on 14 volts, and voltages should be measured in the automatic tuning circuits with a VTVM while the tuner is "seeking." If the signals are weak (check D1 and D2) or the radio is dead when tuning manually, and the automatic tuning system continues to "seek" but does not stop on any stations, go directly to the radio's RF, converter or IF circuits and look for a defect (or defects).

In a radio appears dead while tuning manually but the automatic tuning mechanism stops at different points on the dial, look for trouble in the audio section. The automatic tuning circuit is usually at fault if stations are received by manual tuning but the automatic tuning mechanism fails to stop on

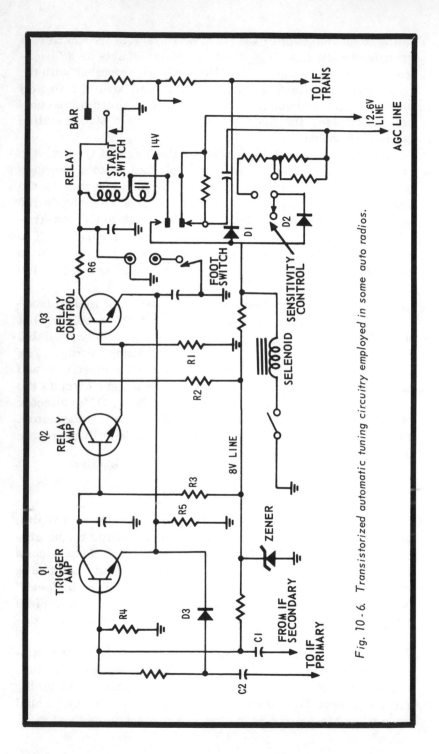

Fig. 10 - 6. Transistorized automatic tuning circuitry employed in some auto radios.

stations and continues seeking. In this case, try switching Q2 "off" by grounding its base momentarily. If the automatic tuning process stops, then check for trouble in Q1's circuit.

If grounding Q2's base does not stop the automatic tuning process, then look for trouble in the Q2 or Q3 circuits. Switch Q3 "off" by shorting the base to the emitter. If the tuning process stops, then Q2, R1, or a resistor in the sensitivity control circuit is probably defective. Also check the start switch.

If grounding Q2's base does stop the automatic tuning process and the voltages around Q1 are not correct, then try this: Connect a VTVM from Q1's base to ground. Watch the meter when the automatic tuner moves the dial pointer through a strong station. If the voltage varies only a small amount, check Q1 on an in-circuit transistor tester; check R4, R5, and the electrolytic capacitor between the collector of the trigger amplifier and base of the relay amplifier in those circuits that use a capacitor there. If the voltage does not change at all when the dial pointer passes through a strong station, then you probably have a very weak signal either at the input or output of Q1. Check the IF alignment, the IF transformer coupling capacitors, C1 and C2, resistors R4, R5, and diode D3.

The automatic circuit is probably defective if the radio works normally but stops working (between stations) as soon as the start switch is released.

If the automatic tuning system does not work at all (refuses to "seek") check the voltage to the system. It should be a minimum of 14v. Check the start switch (including foot switch) and relay contacts. Hold the start button down and measure the voltage across R1. It should not be more nor much less than 1.1v on this particular circuit. Measure the voltage between the base and emitter. It should be 0.6v. If these voltages are within 0.1v, then check R5 and R6. If not within this range, look for trouble in Q2's circuit.

On all automatic tuning circuits, it is important that the antenna, RF, converter, IF, and detector stages be working at high efficiency—to provide maximum signal. Otherwise the automatic tuner will not work properly. On many sets, if a pushbutton is held all the way in while the automatic tuner is "seeking," the fuse may blow.

Watch out for "mixed" transistor types in some mobile radios; that is, where both silicon and germanium transistors are used. This is extremely important when measuring bias voltages and when replacing transistors. When germaniums are used, they will usually be in the power output stages.

When a power transistor is replaced in those sets that use bias pots in the power output collector circuit, remember to reset this pot to the exact voltage recommended by manufacturers' service manuals. Make sure 14v is being applied to the radio when the pot is set.

Never switch a transistor auto radio on to check it without a load across the set's speaker terminals. You can "blow" the power output transistor or transistors. Connect either the proper impedance speaker or load resistor across the terminals before switching the radio on. Do not depend on those speaker sockets that short output terminals when the speaker plug is removed. In fact, this speaker "interlock" sometimes causes trouble by failing to open when the speaker plug is inserted.

Some mobile radio technicians use the "plop" method to check both the speaker and audio output transformer (usually an auto-transformer type) with a VOM. The radio is switched off and the ohms function switch is set to R x 1, the negative meter lead is grounded and the probe lead is touched on-and-off the transistor collector (the case usually).

If a "pop," "thump," or "plop" is heard in the speaker, it is assumed that the output transformer and speaker voice coil are good. But don't depend on it. The aforementioned speaker interlock can be defective and remain closed. Additionally, many radios use an electrolytic bypass capacitor in the power output collector circuit. A shorted capacitor can also prevent a "plop" being heard from the speaker. To be sure, check both output transformer and speaker voice coil for continuity. Check both sides of the autotransformer from the center tap.

As previously indicated, before you remove a radio from its mount, check the antenna and (if the radio is "dead"), check the speaker (or speakers), speaker wiring, and output connections. If a very weak signal or only strong local stations are heard, pay particular attention to determining if the antenna is defective. High resistance leakage between the center wire and the outside shield of the antenna lead-in can be caused by water or moisture accumulation. The only sure way to check this is to remove the antenna. If the antenna proves to be open on a continuity check, don't waste time with it. Install a new one.

You will find quite a few defective speakers—open voice

coils, warped, torn or loose cones—that cause a "dead" radio condition or distorted audio. If the cone is loose around the speaker housing or has a thin crack in it, then use service cement to repair it. Otherwise (and if there are holes in the cone), replace the speaker.

Unless you are thoroughly familiar with the pecularities and know exactly how your in-circuit transistor tester acts on transistors in direct-coupled circuits, disconnect the transistor before attempting to check it. Other basic troubleshooting and repair techniques—defective stage and component isolation procedures—are similar to those previously outlined in Chapters covering solid-state home-type AM and AM/FM radios.

NOISE SUPPRESSION

Although ignition and other noise from auto and inboard boat engines is not the problem it once was, we still have various types of interference that "bug" mobile entertainment equipment. Despite improvements made by engine manufacturers, the inclusion of noise-limiting circuits in receiving equipment —especially FM receivers—and other design improvements, we are still confronted frequently with various noises that interfere with AM, FM and portable all-band radios (also used in autos and boats). This noise includes brake static, intermittent grounds through front wheel bearings and tire static intermittently scraping mechanical connections between parts in the car, hash from the voltage regulator, and generator commutator "whine." A variety of other "static" sources also exist on cars and boats.

If spark plugs do not have suppressors, then recommend suppressor- or resistor-type plugs to your customer. "Pinch-fit" type terminal connectors used in high-tension cables in cars and on boat engines frequently leak and require replacing. Some cables will require replacement with shielded types.

To eliminate generator whine, use a 0.25 mfd (or somewhat larger) interference-eliminator capacitor connected in parallel with a 0.001-mfd mica or general-purpose ceramic capacitor from the terminal to ground. Few "prescriptions" can be written for mobile-type radio interference: it is a "cut-and-try" deal in most cases.

REVERBERATION EQUIPMENT

The idea of reverberation, with equipment providing delay

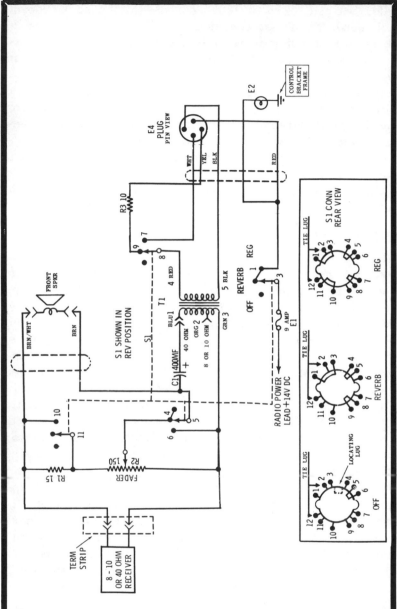

Fig. 10-7. Reverb switching and front speaker circuitry of a Motorola unit.

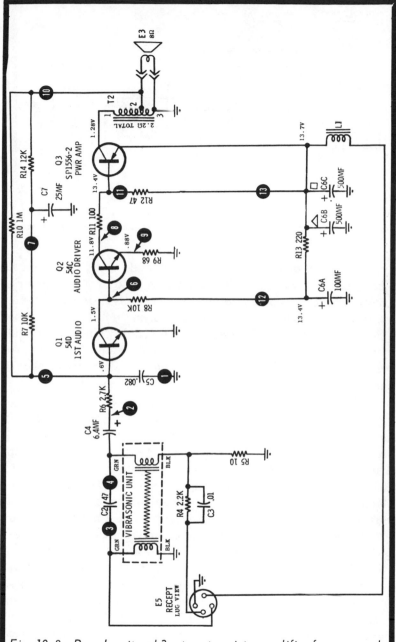

Fig. 10-8. Reverb unit and 3-stage transistor amplifier for rear speaker. (Courtesy Motorola)

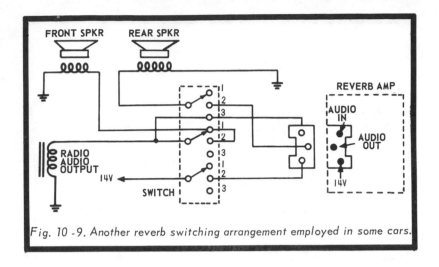

Fig. 10-9. Another reverb switching arrangement employed in some cars.

times from about 20 to 30 microseconds (μs), is to create music effects in the small space of an auto or the interior of a boat which are possible only in medium-to-large listening rooms. Reverb equipment employed on mobile radios consists primarily of an electromechanical delay line, a separate amplifier and a switching arrangement. Mono or stereo reverb is normally applied only on a rear speaker (or speakers). But various switching arrangements are possible.

Two delay-line types are generally employed on mobile radios today. They are basically similar. A switching arrangement, including fader, output transformer, and front speaker is shown in Fig. 10-7. The reverb unit, 3-stage transistor amplifier, output transformer, and rear speaker is shown in Fig. 10-8. Another switching arrangement is shown in Fig. 10-9. Since reverb units are not too efficient, it is necessary to closely match the impedance at the reverb unit input and output.

Many reverb units are installed in auto trunks. If not properly installed, they can develop troubles. And, if engine or other interference is received on the front speakers, it will be present also in the rear reverb speakers. But the longer rear speaker leads can make the noise sound louder in rear speakers. If you get a complaint that switching on or off lights, directional signals, or the horn causes noise in rear speakers, the reverb unit may have an installation defect. At any rate, here again, you may have to apply filters (0.5 to 1 mfd) to whatever unit is causing the noise. Incorrect lead dress in some reverbs can set off oscillations in rear speakers, too.

Some older reverb units use a volume pot at their input. It may be necessary to set this up to provide the effect which a particular customer desires. This is done by first setting the radio's volume control to a "rear-speaker-only" level which satisfies the owner. Then switch to "reverb" and adjust the reverb input pot to a level which the customer approves.

Quite a few older and longer length reverb units may be sensitive to car-body vibration and develop microphonics. If you get a complaint of this type you won't be able to do much about it. Suggest that a shorter, more ruggedly mounted reverb unit be installed.

If you bump into a problem you can't solve, contact your local mobile radio distributor or the factory. They have had experience with almost every possible problem that can arise. FM auto radios are serviced like home-type equipment and the FM receiver described in Chapter 9, as previously stated, was designed for mobile use.

MOBILE TAPE-PLAYERS

Like radios designed for mobile use, mobile tape players are somewhat more ruggedly built than home-type equipment. Otherwise, they are basically similar with a few exceptions, including power supplies, automatic motor-speed control circuits, and an automatic "change-over" switch (from radio to

Fig. 10-10. This Lear Jet Model A-239 combination FM stereo tape-player has a provision for connecting multiplex adapter.

tape player in radio/tape player combinations) which is actuated when the cartridge is inserted. A few other small differences exist.

From an overall view, however, we will be confronted with (1) separate tape players having mounting brackets for individual mounting; (2) combination AM radio/tape players; (3) combination FM stereo/tape players (see Fig. 10-10) and (4) those tape player or tape player/recorder units that operate through existing mobile radios by using an oscillator. At present, you will rarely see anything else except the unitized types which have plug-in AM, FM, and tape-player modules. Depending on the space available, tape players are normally mounted in or under-dash or atop the drive-shaft "hump" on the floor. And there are compact arm-rest types having built-in speakers.

Some tape players are designed to play 4-track pre-recorded cartridges, some 8-track cartridges, and others will play both cartridge types, but in the latter case the 4-track mode may require manual start and channel-selection switching. Most players are designed for negative-ground operation but some are arranged to operate on either negative or positive ground.

Some "crystal-ballers" believe anything can happen in the tape player industry during the next few years, and based on recently past events it probably will. We have already experienced a departure from the conventional front- and rear-seat speaker concept in tape player and FM stereo receiver installations. With all the other extras—including air-conditioners and back-seat TVs—speakers may be mounted under the dash panel, on the kick panels ahead of the front doors, under front or rear seat corners (front outside area), or rear package deck, in front door panels, rear door panels or wherever. And now there are four instead of one or two speakers!

To cut down on the labor and engineering involved, most stereo tape player manufacturers recommend mounting the four speakers in the two front doors, as shown in Fig. 10-11. This is necessary anyway in 2-door cars unless you mount two speakers on the rear package deck. In 4-door cars the best arrangement for stereo effect would be one speaker in each front door and one in each rear door. But this involves considerable work, too. One time-consuming problem is mounting the speakers and routing the wires through the hinged edge of the doors in a way to prevent eventual breakage.

MAINTENANCE AND INSTALLATION PROBLEMS

As stated in Chapter 9 for home-type tape players, your best service approach is preventive maintenance. And the faults that develop in mobile tape players will approximate those that arise in home-type players. It seems clear, however, with only a few years' experience behind us, that mobile-type tape players will require more frequent preventive maintenance. This includes head and capstan cleaning, drive assembly cleaning, head adjustment, and degaussing. Openings in the end of cartridges should be cleaned carefully with a dry, soft brush, or encourage the owner to do so.

Crosstalk between tracks is a frequent problem—especially on 8-track equipment—and this necessitates head adjustment. To adjust 8-track heads properly, pre-recorded test tapes are essential.

It is not necessary to remove a tape player from its mount to clean the head and capstan. As shown in Fig. 10-12, with the tape cartridge removed, the head can be cleaned up using a cotton swab dipped in alcohol (or other manufacturer approved cleaning fluid), and squeezed out. Depress the start switch with the eraser end of a pencil and run the swab up and down the capstan as it turns. Then wipe the head and capstan with a clean, dry swab. All precautions previously listed for home-type tape players and recorders should be carefully observed.

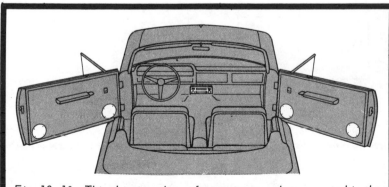

Fig. 10-11. This drawing shows four stereo speakers mounted in the front doors of an auto. (Courtesy Lear Jet)

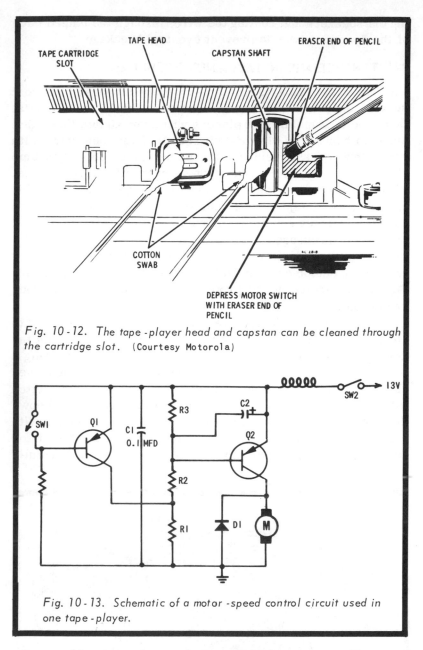

Fig. 10-12. The tape-player head and capstan can be cleaned through the cartridge slot. (Courtesy Motorola)

Fig. 10-13. Schematic of a motor-speed control circuit used in one tape-player.

Some older tape players have polished capstans. If you get one for repair and the tape slips while playing, roughen the capstan lightly with fine sandpaper while it is turning. Later models of this same tape player have pre-roughened capstans.

A number of motor-speed control circuits are employed in tape players, but we will look at only one circuit here. This type employs a mechanical governor, a 3000 RPM 4-lead motor and a 2-transistor regulator circuit. The circuit is designed to switch the motor on and off between 100 and 200 times a second to stabilize its speed. In a simplified way, the circuit works as follows (see Fig. 10-13):

When a tape cartridge is inserted in the player, SW2 closes and 13 volts is applied to Q2's emitter. In effect, Q2 operates as a variable impedance in series with the motor (M). The impedance varies with Q2's bias, maintaining 7 to 8v on the motor through its collector. While Q2 conducts, SW1 remains closed when the motor speed is below 3000 RPM. Q1 is cut off.

When the motor speed increases above 3000 RPM, the centrifugal force opens SW1, Q1 conducts heavily through R1 to ground, biasing Q2 off through R2 and power to the motor is switched off. D1, shunted across the motor, filters transients caused by rapid motor starts and stops. C1 eliminates decay noise created when the motor slows after a cartridge is removed. C2, a large capacity electrolytic, aids in maintaining bias stability.

You will need individual service manuals for tape players, combination AM radio/tape players, FM radio/tape players and the combination type which uses AM, FM radio and tape player plug-in modules.

CHAPTER 11

Servicing Automatic
Record Players

Like tape recorders and tape players, record players and manual turntables are electromechanical devices. And with the exception of a relatively few design variations, most record players are similar and more standardized than tape recorders or tape players. The number of major manufacturers at the present time probably does not exceed eight or ten, and chances are there will be a smaller number next year.

After you work on a few dozen record players you will begin to recognize the various OEM "crop marks," and when you run into a similar unit again you soon find that you have become familiar with all of them. From the servicing viewpoint, the best way to approach record players is the same as we do with all other electronic and electromechanical equipment: Break the equipment down into specific sections, and then quickly learn what symptoms are peculiar to each section.

The first section we are concerned with is the drive system. This includes the motor, drivewheel (or idler), motor pulley, and speed-regulating components. Some high-class turntables use belt drives, thus minimizing "rumble."

The cycling sections of record changers are perhaps the most confusing. Here, we are concerned with start, stop, record-drop, shut-off, indexing, anti-skating, and change-cycle components and functions. This section also includes "manual operation" and trip-mechanism components and their functions. All we have left are tone arms, cartridges, and styli (needles).

Let's select at random one modern automatic record player

and go over its various sections briefly. A photograph of the unit (without spindle attached) is shown in Fig. 11-1. (All service information and illustrations were supplied by United Audio Products, unless otherwise noted. The numbers in parentheses refer to the "call-out" numbers in the illustrations.)

We'll look at the motor and drive section first. The motor suspension and turntable drive system is shown in Fig. 11-2. The turntable and change-cycle components are driven by a four-pole induction motor (116). Replaceable motor pulleys (105) adapt the player to either 50- or 60-Hz power line operation.

The drive wheel (90) drives the turntable, and when the tone arm is in the rest position the wheel disengages to prevent it from becoming deformed because of pressure. It should be noted at this point that you may run into a variation of this arrangement. There may be two drive wheels instead of one. But don't let this bother you; the particular manufacturer's service literature will give operating details.

When the speed control knob (8) is switched to either 16-2/3, 33-1/3, 45, or 78 RPM, the drive wheel moves up or down to engage the proper step on the motor pulley. This takes

Fig. 11-1. Typical high-grade automatic record player.

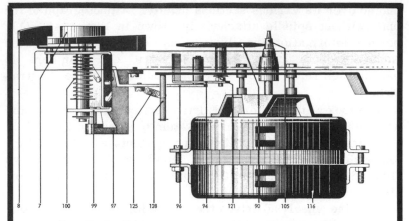

Fig. 11 -2. Motor suspension and turntable drive system.

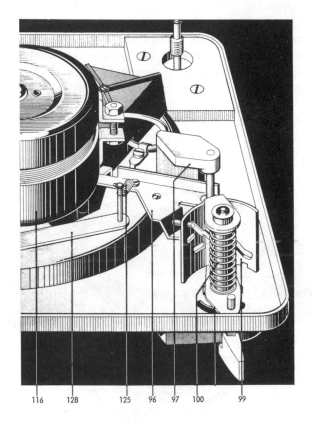

Fig. 11 -3. Turntable speed and drive -wheel shift mechanism.

place through rotation of the switch segment (99) which causes the rocker assembly (96) to move up and down, in turn moving the drive wheel up or down. A fine-speed regulating system provides an approximate 6% variation for all four speeds. Referring to Figs. 11-2 and 11-3, when control knob (7) is varied, switch segment (99) causes the rocker assembly (96) to move up and down. This small vertical motion forces the drive wheel to a slightly different position on the motor pulley. The tapered shape of the pulley provides a $\pm 3\%$ variation each side of the nominal speed selected by the speed-change knob (8).

Trouble symptoms in this area include: (1) Turntable does not run; (2) turntable does not run at proper speed; (3) correct speed is obtained only by extreme adjustment of fine-speed regulator; (4) fine-speed regulator does not operate. The causes for (1) include no power to the motor, the drive wheel is not in contact with the turntable, or the motor pulley is loose. The causes for (2) include an incorrect motor pulley for the local line frequency, (3) the drive wheel is slipping against the motor pulley, or (4) friction in the motor bearing. Symptom (3) is caused by improper drive wheel contact with the motor pulley, and symptom (4) by the control knob (7) being pushed down by rough treatment or during shipment.

CYCLING

The start of the operating cycle begins when switch (82b), Fig. 11-4, is placed in the start position. This moves the switch lever (233), as shown in Fig. 11-5, toward the main cam, which initiates the following sequence:

1. The switch lever assembly set screw (184) turns the switch arm (128) mounted on the grooved shaft (182). As previously indicated (Fig. 11-2), the drive wheel (90) is moved into contact with the motor pulley and turntable by the rocker assembly (96) and a tension spring. Simultaneously, the switch slide (118), shown in Fig. 11-5, and the switch arm actuate the line switch, starting the turntable.

2. Referring to Fig. 11-4, the switch angle mounted on the switch lever assembly (233) is brought within range of the cam-follower lever (137), so it is pushed into the change position after rotation of the main cam.

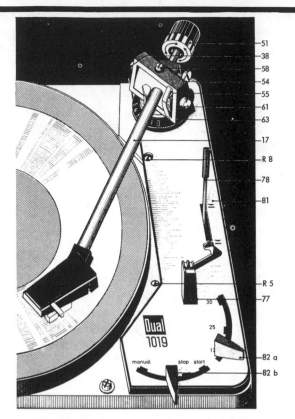

Fig. 11 -4. Operating and control elements.

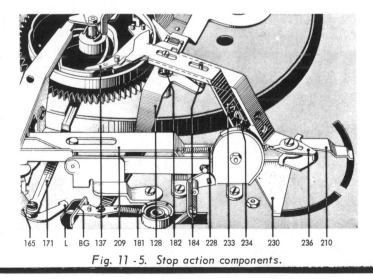

Fig. 11 -5. Stop action components.

When the operating switch is moved, it also releases the start lever (236) which is pulled toward the main cam by a tension spring (234). This causes the coiled spring (177), Fig. 11-5, to bring the shut-off lever (131), Fig. 11-7, within range of the main cam "dog." The shut-off lever drives the main cam. The operating switch is locked during the start cycle (when the main cam is turning). Just before the main cam reaches its null position—at the end of the change cycle —the start lever is pushed clear of the main cam by the start pin (SB), Fig. 11-6. This restores the switch lever and operating switch to their original positions.

When installing a new record changer or after moving one, the changer should be started while the tone arm is locked. This will automatically readjust the shut-off lever which may have shifted out of position. Placing the operating switch in the "manual" position also begins the start cycle as previously explained, but in this position, when the turntable begins to rotate, the switch arm latch (179), Fig. 11-8, rests in the support (BG), Fig. 11-5, locking the switch arm in position to keep the drive wheel in contact with the turntable.

When the tone arm reaches the shut-off groove, it returns to its rest position and the player is shut off. If the tone arm is lifted off manually, however, and returned to its rest, the

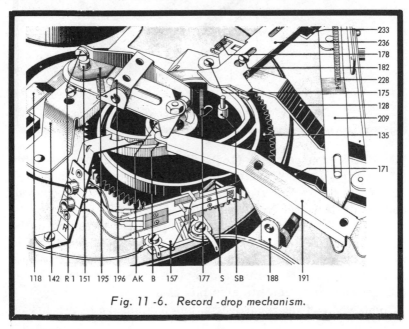

Fig. 11-6. Record-drop mechanism.

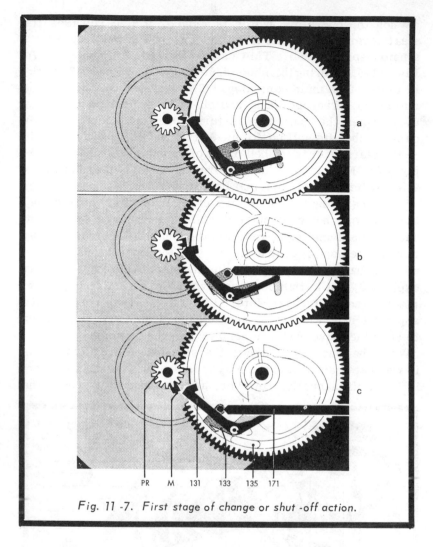

PR   M   131   133   135   171

Fig. 11 -7. First stage of change or shut -off action.

tabs of the arm segment (165), Fig. 11-5, release the latch (179), Fig. 11-8. Tension spring (181) returns the switch arm (128) to its initial position, the line switch is opened, and the drive wheel is disengaged.

Referring to Fig. 11-5, when the operating switch is in the "stop" position the switch lever (233) and switch angle move toward the main cam, as in the start cycle, but only half as far. This causes the main cam to push the cam-follower lever (137) to the side, into its stop position.

Components of the record-drop mechanism are shown in Fig.

11-6. This record player uses two spindles, one for standard records and one for 45 RPM records. The main cam is designed so a record can drop only when the tone arm is above its rest—at a point where it cannot impede the largest 12" diameter record. The dropping process takes place through rotation of the main cam (135) and cam (AK) which guides the cam rocker (196), pushing the change actuator stud (151) which releases a record through action of the spindle mechanism.

A muting switch (157) is used to prevent the tone arm cartridge from picking up change-cycle noises. Switch springs for both channels are actuated by the main cam (135). The muting switch is open in the rest position.

The position of the cam-follower lever (137), Fig. 11-9, determines shut-off and change functions. When the last record drops from the spindle, the change lever guides the cam-follower lever (137).

Shut-off is initiated when the change lever brings the cam-follower lever into position; the longer end is brought toward the center of the main cam. After the tone arm swings over its rest, the guide post (B), Fig. 11-6, on the main lever (191) contacts the outside of the main cam (135) whose vertical profile causes the tone arm to lower onto its support. Traversing of the tone arm releases the latch (179) from its support (BG). The main cam keeps the switch arm (128) in

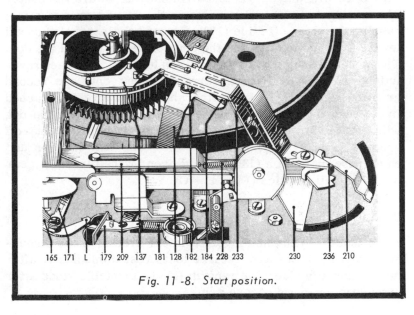

165 171 L   179 209 137 181 128 182 184 228 233   230   236 210

Fig. 11-8. Start position.

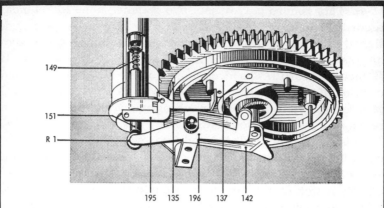

Fig. 11-9. Shut-off position. The position of the cam-follower lever selects shut-off and change functions.

its "play" position until the end of the change cycle. When the main cam returns to its null position, the switch arm drops into the cut out in the main cam, the line switch is operated, and the drive wheel is once again disengaged.

Referring to Fig. 11-7, the "dog" (M) on the turntable gear (PR) and the shut-off lever (131) actuate both the change cycle at the end of a record and shut-off after the last record is played.

As a record plays, the tone arm moves toward the center of the record at a rate dependent on the record-groove pitch, pushing the shut-off slide (171) forward against the shut-off lever. This motion carries the shut-off lever toward the dog. The eccentric dog pushes the shut-off lever back at each revolution, so long as the tone arm advances only one record groove (Fig. 11-7A). When the shut-off groove is reached in the record, the shut-off lever is brought against the dog with greater force caused by the greater pitch of the shut-off grooves (Fig. 11-7B). The shut-off lever then engages and causes the main cam (135) to be driven by the turntable gear out of its null position (Fig. 11-7C).

CYCLING PROBLEMS

The trouble symptoms involved in the cycling areas discussed include (1) tone arm returns to its rest immediately after being manually placed on record; (2) the turntable stops after automatic set-down of tone arm.

Symptom number 1 is caused, usually, by rough treatment of the shut-off mechanism. The cure is to put the player through the "start" cycle as previously mentioned. Number 2 is caused by failure of the switch arm (128) to engage latch (179), Fig. 11-10. Loosen the screw (175) and turn the short arm piece on the long switch arm piece. Turn the main cam to its null position and adjust for about 1/64" play between the tabs (L), Fig. 11-8, and the arm segment when the tone arm drops into its rest.

Other trouble symptoms include (3) the tone arm returns to its rest position after each record is played. (4) Turntable does not turn when the switch is moved to "manual" and the tone arm is off its resting post. (5) The last record continues to repeat play. (6) A record drops after the switch is moved to "stop" and another record drops when the switch is moved to "start." (7) Records do not drop. (8) Turntable slows down as record drops. (9) Acoustic feedback.

Symptom number 3 is caused by excessive engagement between change lever (195), Fig. 11-9. With a record on and the spindle locked, the change lever should be readjusted so a 1/64" clearance exists between the change lever and the guide post of the cam-follower lever (137). With no record loaded, engagement should be about 1/32" to obtain shut-off.

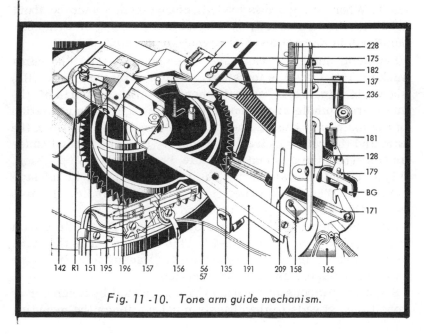

Fig. 11-10. Tone arm guide mechanism.

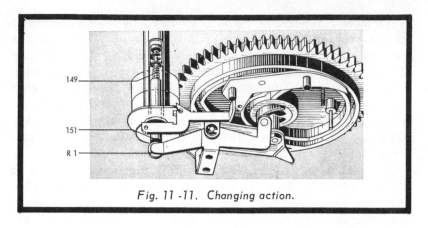

Fig. 11 -11. Changing action.

The cause of symptom number 4 is a misadjusted switch lever assembly. With the switch in the "manual" position, setscrew (184) is readjusted so the latch (179) overtravels the support (BG) about 1/64". Make sure the adjustment is secured by the locknut (see Figs. 11-5 and 11-8).

Symptom number 5 is caused by inadequate engagement between change lever (195) and cam-follower (137). Referring to Fig. 11-9, with a record on and the spindle locked, readjust change lever (195) to obtain about 1/64" clearance between the change lever and the guide-in of the cam-follower lever (137). When no record is loaded, engagement should be about 1/32" to obtain shut-off. The situation existing in symptom number 6 is normal and you cannot do anything about it.

Symptom number 7 can have three causes. (1) The cam-rocker (196) travel is too short. Readjust the eccentric (R1) shown in Fig. 11-11 so that when the three supports of the automatic spindle are completely retracted, further rotation of the main cam causes overtravel of about 1/64" between the cam and the roller of the cam rocker. (2) The second cause arises when the automatic spindle is not locked in position. When the spindle is inserted, it must be rotated to its stop point. (3) The third cause is a defective spindle. In this case, the spindle must be replaced.

Symptom number 8 is caused by too-long travel of the cam rocker (196) seen in Fig. 11-7. This adjustment is the same as for cause 1 in symptom number 7. Eccentric R1 must be properly adjusted as previously stated.

Symptom 9 (acoustic feedback) can be caused by a number of conditions. Leads to the chassis must not touch the mounting board. Cut-outs should be changed according to installation instructions. When connecting leads are pulled too tight, loosen or lengthen the leads. It should be understood here that the suspension system of a turntable must be properly designed to minimize acoustic feedback. High-level volume from speakers can cause the floor, ceiling, walls, and furniture in a room to vibrate. Precautions must be taken to prevent these vibrations from being transferred to the stylus. Under certain circumstances the record player may have to be placed farther away from the speakers.

TONE ARM

Although tone arms are more ruggedly built than they were a few years ago, they probably give us more trouble than any other record player section. A tone arm sticks out like a "sore thumb," it's in constant use and comes in for considerable mishandling and abuse.

A tone arm and its bearing assembly is shown in Fig. 11-12. An underneath view is shown in Fig. 11-13. Its suspension with anti-skating compensation is shown in Fig. 11-14. In servicing, both bearings in the tone arm bearing assembly require a small, barely noticeable amount of play. Only the left bearing screw (38) is used for adjustment of the horizontal

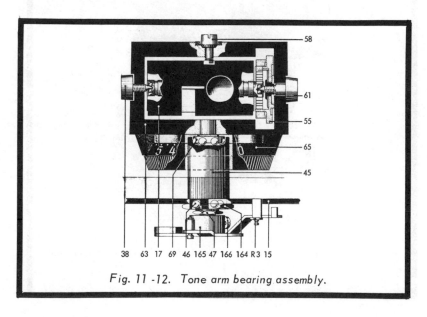

Fig. 11-12. Tone arm bearing assembly.

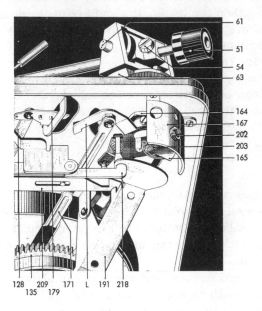

Fig. 11-13. Underneath view of tone arm suspension.

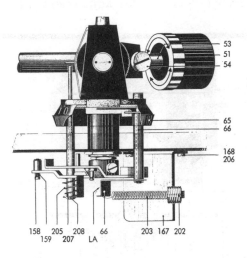

Fig. 11-14. Tone arm suspension with anti-skating compensation.

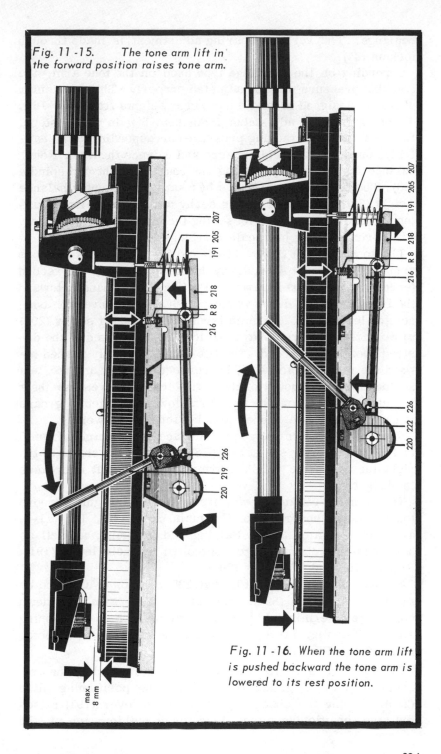

Fig. 11-15.   The tone arm lift in the forward position raises tone arm.

207
205
191
218
R 8
216

226
219 220

max.
8 mm

207
205
191
218
R 8
216

226
222
220

Fig. 11-16. When the tone arm lift is pushed backward the tone arm is lowered to its rest position.

bearings. The vertical bearing adjustment is made through locknut (47).

Depending on the cartridge type used on the tone arm, the tracking pressure must be adjusted properly. This particular player is designed to handle cartridge weights from 1 to 16 g. Tracking force is adjustable from 0 to 5 g in 1/2-g steps. Before setting the tracking pressure corresponding to the cartridge used, set the scale to zero and balance the arm. Rough balance is obtained by sliding the counterweight and spindle (53). The final setting is made by rotating the counter balance weight (51). Tracking force is set by turning the spring housing (55) and tightening or loosening the internal spiral spring. The spring housing has scale markings in 1/2-g steps.

The tone arm lift (Figs. 11-15 and 11-16) permits the tone arm to be safely set down at any desired position of the record (except in the shut-off area). Pulling the lift handle toward the front turns the drive washer (226). This moves the connecting lever (218), the main lever (191), and lift screw (207) to raise the tone arm. After the tone arm is moved to the desired spot on the record, the lift handle is lightly pushed toward the rear to release. The connecting lever and the leaf spring (192, not shown) of the main lever (191) resume their normal positions and the tone arm lowers. Silicone grease is used on the drive washer to delay tone arm lowering.

Needle height from the record surface can be varied from 0 to 1/4" by adjusting set-screw (R8). Turning the screw to the right increases the height, turning it to the left decreases the height.

Tone arm lift-off and set-down is controlled by a guide groove located on the underside of the main cam (135), Fig. 11-7, as the cam rotates 360°. Raising and lowering, as well as horizontal movements, are controlled by main lever (191), Fig. 11-16, and lift screw (207).

Accommodation for 7, 10, and 12" records is afforded by the indexing switch (82a), Fig. 11-4. Tone arm set-down points are determined by the eccentric of the arm-positioning slide (209), Fig. 11-10, contacting the record-size selector lever (230), Fig. 11-8.

The tone arm's horizontal movement is limited by the arm segment (165), Fig. 11-10, striking the positioning slide (209). During the change cycle the main lever (191) raises the arm-positioning slide, bringing it within reach of the spring stud (158), Fig. 11-10. When the change cycle is completed—when the tone arm is set down on the record—the arm-positioning slide is again released and returns to its normal position, out of the reach of spring stud (158), which permits unhindered horizontal movement of the tone arm while playing a record.

The anti-skating mechanism shown in Fig. 11-17, eliminates "needle-scratch" of records by preventing the tone arm's "skidding" across the record. Rotating the adjustment ring of the anti-skating mechanism moves the spring lever (66), Fig. 11-13, through the curved track. The curved track inside the adjustment ring moves the spring lever (66) when the anti-skating mechanism adjustment ring is rotated. The ten-

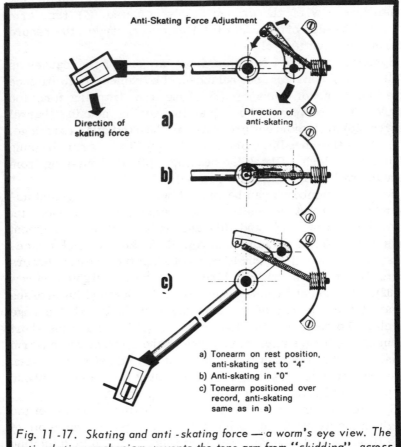

Anti-Skating Force Adjustment

Direction of skating force

a)

Direction of anti-skating

b)

c)

a) Tonearm on rest position, anti-skating set to "4"
b) Anti-skating in "0"
c) Tonearm positioned over record, anti-skating same as in a)

Fig. 11-17. Skating and anti-skating force — a worm's eye view. The anti-skating mechanism prevents the tone arm from "skidding" across records.

sion spring (203) transmits the counter-movement to the tone arm. Optimum adjustment of the anti-skating mechanism is obtained with a needle curvature of 0.7 ($\pm$0.1 mil). The adjusting screw (threaded bushing) is sealed with Glyptol after setting. A "Skate-O-Meter" and standard test record, L096, are required for readjustment and should be performed only by authorized service stations. Separate instructions are available from the manufacturer.

Symptoms of troubles in the tone arm section include: (1) Needle slides out of record groove; (2) tone arm lowers beside tone arm rest; (3) horizontal bearing friction too high; (4) vertical movement of the tone arm is impeded during the set-down cycle; (5) tone arm does not move onto record when the drop cycle is actuated; (6) tone arm lowers too quickly onto record when the drop-cycle is actuated; (7) tone arm misses the edge of the record; (8) tone arm strikes the record during the change cycle.

Symptom number 1 can stem from any one of five causes as follows: (1) Tone arm not balanced. The remedy is to balance the tone arm, of course. (2) Tone arm tracking force too light. The remedy is to adjust the spring housing (barrel screw 55) to the correct pressure for the particular cartridge. (3) Defective needle. Replace it. (4) Tone arm bearing friction too high. Check the bearing. (5) Ball missing from shutoff rail (171). Install a new ball.

Symptom number 2 can be caused by: (1) Arm segment assembly (165) out of position. To remedy this, loosen the machine screws (162 and 166) and rotate the arm segment assembly. Then tighten the screw (166) and recheck the adjustment. The adjustment is correct when the tone arm lowers onto the arm rest without binding. Finally, tighten screw (162). The second cause occurs when the latch (179) presses against the tabs (L) of the arm segment during the change cycle. To remedy this, loosen screw (175). Turn the short arm on the long switch-arm piece to the correct switch-arm position. Turn the main cam by hand and adjust so that when the tone arm lowers onto the arm rest a clearance of about 1/64" should exist between the latch segment tabs.

Symptom 3 can have but one cause: The tone arm is set too high on the arm lift. Main lever jams against the guide pin of the lift-screw assembly. Correct this condition by adjusting screw (R8), Fig. 11-4. This happens when the needle is higher than 1/4" above the record.

Symptom 4 may be the consequence of either of two causes: (1) Bearing friction is too high. Check bearing screw (38) and arm balance. (2) Lift the screw jams in the guide sleeve of the arm segment (165). Correct by removing and cleaning the lift screw.

Symptom 5 indicates that damping is too great, probably caused by a dirty drive washer. To cure this malfunction, loosen nut (224), remove cover washer (225) and drive washer (226). Clean thoroughly. Spread silicone grease evenly on both sides of the drive washer. Reassemble and wipe off excess grease. Symptom 6, caused by too little damping, receives the same treatment as symptom 5.

Symptom 7 may be the result of any of four conditions: (1) Wrong record size selected; (2) set-down incorrectly adjusted; (3) record not standard size; (4) tone arm clutch surfaces contaminated. The remedy for the first cause, of course, is to select the correct record size with the record indexing switch. To remedy the second fault, adjust for 7" records by turning the eccentric screw (R5) so that the tone arm sets down about 1/16" from the record edge. Once the 7" adjustment is correct, the player is automatically correct for 10 and 12" records. For the third malfunction use standard records, obviously, and for the fourth, clean the clutch surfaces.

Symptom 8 is caused by incorrect tone arm height. Adjust the arm height with screw (58). When correctly adjusted, the pickup needle is 1/64" above the dress-up plate (81) when removed from the arm rest. For bottom chassis views, exploded top- and bottom-chassis views of parts, parts lists, and other important service data, see the manufacturer's service manual.

The instructions outlined here, while dealing with a single record player made by one manufacturer, should acquaint you generally with most other players, especially in the areas of cycling, tone arm adjustment, record dropping, indexing, set-down, tone arm height, stylus pressure and other record player sections. But you will run into other problems with somewhat different record player types.

For example, in those units that employ a record pressure arm, variously called over-arm, control arm or "stabilizer," considerable trouble will be experienced primarily because the arm is abused by many owners. Repeated lifting at the front

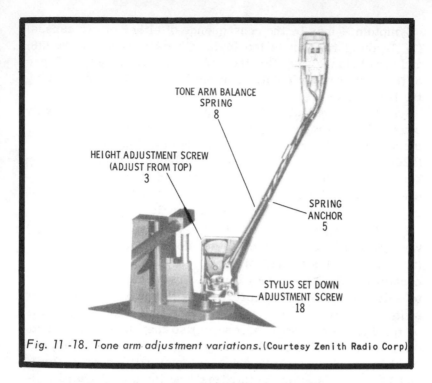

TONE ARM BALANCE
SPRING
8

HEIGHT ADJUSTMENT SCREW
(ADJUST FROM TOP)
3

SPRING
ANCHOR
5

STYLUS SET DOWN
ADJUSTMENT SCREW
18

Fig. 11 -18. Tone arm adjustment variations. (Courtesy Zenith Radio Corp)

end instead of at the rear of the arm will result in springing the arm and its associated shut-off mechanism. Always advise record player owners to grasp and lift this arm at the rear, rather than at the front end.

Other record changer design variations will be noted also in the area of stylus set-down, stylus pressure, tone arm height adjustment and some other areas. One variation is shown in Fig. 11-18. Adjustment screw (18) determines the proper stylus set-down position. Normally, this is adjusted for a 12" record, and when done properly the stylus set-down position is correct for 7 and 10" records. Tone arm height is properly adjusted by screw (3). This is normally done by placing a stack of six average thickness records on the turntable and adjusting the screw until the stylus clears the top record by 1/8".

Proper stylus pressure is obtained by repositioning the balance spring (8) to a higher or lower spring anchor. This pressure will vary from one type or brand of record player to another, depending primarily on the cartridge weight. Check the manufacturer's service data for proper pressure.

GENERAL SERVICE CONSIDERATIONS

In addition to a special work-bench area in your shop, you will need (1) a sturdy heavy adjustable steel support or stand to hold record changers in various positions to facilitate top- and under - chassis adjustments and checks; (2) a complete supply of manufacturers' service manuals; (3) a supply of most used original replacement parts, especially drive wheels, an assortment of "C" washers, "E" rings, springs, regular washers, tone arm wire, AC power cords, etc.; (4) special tools, including a bending tool, set of jeweler's screwdrivers, strobe disc, test records, mil-type feeler gauges, hypo needle (for oiling bearings), tweezers, spring hooks, stylus pressure gauge, gram scales (for springs), styli microscope, etc. The feeler gauges are especially necessary if you are working with tape recorders and tape players, too. You will also find it very convenient to stock a few tone arms, spindles, cartridges, and a good assortment of needles— especially after you discover which types are most in demand as replacements. This comes only after a certain amount of experience, of course.

CLEANERS AND LUBRICANTS

Additionally, a supply of cleaners and lubricants—especially those specifically recommended by record player OEMs—is a must. Trichlorethylene, any of the various naphthas and denatured (isoprophyl) alcohol are usually recommended as cleaners. Carbon tetrachloride (carbontet) should never be used as a cleaner for previously mentioned reasons. Low-lint content paper toweling is helpful. Watch out for cleaners that chemically react with plastic surfaces. Petroleum-base lubricants will also discolor, erode, or distort some plastics.

Most OEMs recommend five or six different "grades" and types of lubricants. For example, a fine oil is normally recommended for bearings. And various other lubricants, like Molycote paste G, silicone grease and thicker, non-gumming oil, Calypsol WIK 700 and others, are recommended—all especially selected to provide optimum results in various record player functional sections.

It should be carefully noted, however, that lubricants should be kept away from all drive surfaces, the inside rim of turntables (on rim-drive systems), belts, clutch levers, etc. It is easy to remember that the incorrect lubricant or a lubricant in the wrong place can easily, and frequently does, create trouble symptoms indistinguishable from those caused by mechanical misadjustment. The experienced technician will tell you: Clean thoroughly, lubricate properly, and then make whatever adjustments are necessary—in that order.

Many changers have springs which allow the changer to "float" on its base-board assembly. Screws protrude through the coiled springs and allow these to be adjusted for proper floating action or for leveling the changer. These screws are normally tightened when the unit is shipped and should be loosened when the record changer is installed. Leveling screws may also be provided on some units. These should be adjusted while using a small carpenter's level to determine when the record changer is perfectly level.

STYLI

A few important things to remember about phono needles include size or diameter and pressure of the needle against the record groove. A 33 1/3 LP mono needle is normally 1-mil (0.001") in diameter. And since a stereo record groove is smaller, the needle should be approximately 0.7-mil (0.0007") in diameter. If a mono stylus is used on stereo records, the records will be damaged. A stereo stylus, however, can be used on a mono record. So, if mixed records are used, that is, both stereo and mono records, a 0.7-mil stylus should always be employed instead of a 1-mil stylus.

And less pressure should be used on stereo styli than on mono styli. This is true because (1) a smaller diameter stylus will wear the record faster if mono stylus pressure is used; more surface noise will be generated also; (2) the vertical compliance necessary to properly operate the stereo cartridge will be dampened if the stylus has too much pressure.

Many high-class record players use elliptical-shaped styli instead of "round" or conical-shaped styli. This should be carefully noted before a needle is replaced in a cartridge. And this stylus must be accurately replaced by positioning it correctly on the cantilever suspension so that its width-axis (0.0007") is square to the cartridge body.

A final suggestion: An important element in your customer-relations stance includes customer education. In this particular case, care of record changer equipment—especially expensive records. Investing a little time and a few dollars here will pay future dividends. Purchase a few hundred booklets on record care and give them to your Hi Fi customers. One such booklet, "How to Clean, Maintain and Protect Records," is available from Elpa Marketing Industries, Inc., New Hyde Park, N.Y. 11040.

CHAPTER 12
Selling and Installing
Audio Systems

More people become interested in good music every year. And, despite the upsurge of audio component sales by chain store and mail-order operators, plus competition from the intermittent popularity of combination "Hi Fi" packages, more component-type home music systems are being sold and installed by service-dealers each year. Moreover, a number of independent surveys show that the bulk of component sales and installations are made in small and medium size towns and cities and in suburban areas of large cities. It also has been clearly indicated that both audio specialists and general TV-radio-Hi Fi dealers are participating successfully in this sales growth.

SALES / SERVICE APPROACH

In a high percentage of those cases reviewed, the most successful operators proved to be those who had good sales-demonstration facilities and were well equipped to provide top grade installation and maintenance services. Local FM radio-station advertising ranked high as an important factor in promoting sales. But, more important was the revelation that sales and service are interdependent. Good initial customer relations, good installation practices and subsequent reliable service definitely increased sales through word-of-mouth advertising—the increased sales ultimately expanding service revenues. No matter how you view it, good service is a powerful sales promoter.

In many cases which the author has personally investigated,

successful audio-component sales began with a properly designed and acoustically-treated "listening" room at the sales-service location. You will also need an average size home "listening" room mock-up with properly spaced speakers and comfortable chairs at the focal listening point. Various types of equipment—an AM/FM radio tuner, record player, and tape player—should be easily switchable from a central control unit into stereo preamps and amplifiers.

The next requirement is an intelligent customer-relations program, plus a thorough knowledge of good installation practices which begins, logically, with a home survey. You must see the home "listening" room which the prospective customer has in mind. The idea is to provide listeners with the most satisfaction for the money spent—come as close as you can to providing "concert-hall realism."

Because of the general public's lack of technical know-how and the "bafflegab" contained in do-it-yourselfer-oriented advertising, you are obliged to help prospective customers select the equipment best suited to provide maximum satisfaction under individual circumstances. And you have to explode the bubble-concept of "stereo" which exists in the minds of many prospective customers. Whether you advise the prospective customer what equipment to buy or you sell him what he asks for, he will blame you if he is not satisfied. Hence, it is better to make an effort to protect your long-range interests by catering to the prospect's immediate interests. From the long-range business viewpoint, and considering the nature of Hi Fi equipment, you also have a better chance to "sell-up"—accept a trade-in at some future date—provide the customer with more elaborate equipment when his desires and tastes become more discriminating.

Another important point should be emphasized here: Sell state-of-the-art equipment. Avoid wasting your time with so-called "Hi Fi portable" equipment which is in the "toy" class, whether phonographs, tape recorders, or whatever. Let the discount houses, drugstores, etc., have the headaches. The fast buck you may lose on the "teen" market won't hurt you in the long run. Don't waste your time with the so-called Hi Fi "buffs" who ask you to "design" something that hasn't been made yet. And, if someone wants a "center" (or third stereo channel), suggest a third amplifier, speaker, and a mixer system. But make it clear that you do not guarantee satisfaction in a case like this. In fact, it is doubtful that a "center" channel will improve a stereo system except in a large listening room where the two regular speakers are spaced very widely apart. After all, there are only two channels recorded on phonograph records and tape, only two channels broadcast on FM stereo!

THE HOME "LISTENING" ROOM

Although it is beyond the scope of this Chapter to deal at length with the acoustical properties of various room sizes, reflecting and absorbing characteristics of building, furnishing, and decorative materials, it is necessary to scout the subject briefly.

Why is it important to see the room in which the customer wishes to install his Hi Fi components? Simply because you cannot really serve the prospect intelligently without first knowing the room size, height of ceiling, whether the room floor is bare, has wall-to-wall carpeting, or is partly covered by one small rug or scatter rugs; the kind of walls—wood or painted plaster; whether walls have decorative rugs and how many; number of windows, and whether they are draped or bare; how much upholstered furniture is in the room and a few other bits of information.

We want to advise the prospect how to get the most satisfaction from his audio equipment. And we do not want to install equipment in either a "dead" sound-proofed room or in a room that's as "alive" as a marble cave or a large tiled bathroom.

Although room dimensions are extremely important in obtaining the best music reproduction, especially in large area rooms, there's nothing we can do about altering the room physically, except in a limited way. We can, however, subdue undesirable room resonances and tame reverberation by proper acoustic treatment. And if we run into some problems we can do nothing about—especially in the low-frequency areas—the customer can be instructed how to artifically compensate somewhat for this by properly adjusting the equipment controls.

We will seldom have problems with the average small room, unless it is a "cubicle" like one we ran into in a 60-year-old house which was about 10' x 10' with a 9'-8" high ceiling. The thick plaster walls and ceiling had been painted over many times with hard-gloss paint. But we must try, for example, to avoid installing speakers against similarly treated walls. That is to say, the wall opposite the speakers should be reasonably sound absorbent if the speaker wall is highly reflecting, and vice-versa. This, however, is only a rule-of-thumb observation. If you want to go deeper into this subject, specialized books cover the gamut of room-size characteristics and acoustical properties of various building and interior decorating materials and how they can be used to alter undesirable room characteristics.

Suppose, for example, you were asked to "survey" a 12 x 18' room having an 8' ceiling. (Most ceilings in present-day apartments and homes are slightly lower.) You find the walls and ceiling are heavily painted plaster and a medium-length upholstered couch covers an unused fireplace along one 12' wall. The two windows on one 18' wall have shades and curtains but no drapes. The 12' wall opposite the fireplace has 6' wide open archway leading to another room. Furniture includes two upholstered chairs, and three small scatter-rugs partially cover a hardwood floor.

First, you may decide not to disturb the room arrangement and try speakers on each side of the open archway, inviting the prospect to listen to various types of recorded music for about a half hour. Then, after a quick trip to your shop, you may place a sheet of 1/4" plywood temporarily over the archway opening, hang a large piece of drape material temporarily over the two windows, unroll an 8 x 12-ft rug on the floor, rearrange the speakers, couch, and chairs as shown in Fig. 12-1. Again, you invite the prospect to listen to the same half-hour selection of recorded music. They would no doubt like this arrangement much better.

Perhaps this family will buy not only about $1500 worth of audio equipment (and pay 10 to $15 an hour for your time), but also an accordian-type sliding door for the open archway, a 9 x 12-ft rug for the floor and two draw-string drapes to cover both windows and the space in between, plus a few other items.

For example, you may have suggested a 3 x 4 1/2' wall rug between the speakers and 18 x 24' paintings to hang above each speaker. You might also have raised the medium-size speakers by placing them on wooden bases, thus the center of the speakers probably would be about 30 to 34" from the floor,

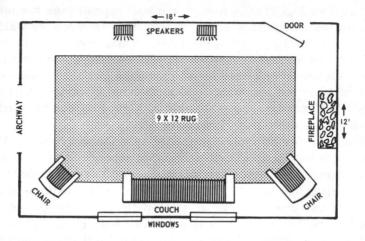

Fig. 12-1. Basic arrangement of a typical medium-size home listening system.

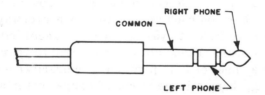

Fig. 12-2. Stereo phone plug.

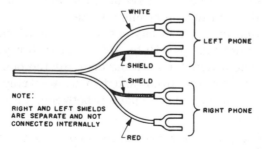

NOTE:

RIGHT AND LEFT SHIELDS ARE SEPARATE AND NOT CONNECTED INTERNALLY

Fig. 12-3. Stereo headphone lug-type terminations.

or at approximately the average ear level of listeners seated on the couch and in the chairs.

You would, without doubt, have tried different speaker spacings along the 12' wall—from about 4' up to 8'. And the listeners probably would have said that they noticed little or no difference in this spacing, but insist that a 6' to 7' spacing "sounded best."

Finally, in addition to the two speakers, you probably ended up by installing an AM/FM tuner, a separate preamp/control unit, a 40w-per-channel stereo amplifier, an automatic turntable and a cartridge-type tape player—all mounted on room-divider-type shelves in one corner of the room next to a chair. Or, you may have decided, with the prospect's permission, to mount all this equipment in a specially selected wood cabinet instead of shelves, placing the cabinet along side the same easy chair.

Because of the "generation gap" existing in most families today—and this is perhaps more obvious in musical tastes— you are in a position to sell and install two separate music systems in most homes. The one for the younger members of the household will probably be less elaborate but it should not be in the "toy" class. The "compacts," employing an omnidirectional speaker system or two speaker units, are frequently ideal. The omnidirectional speaker system works well if it is to be placed at the center of a rumpus room. Compacts come in AM/FM radio, phono, and tape player combinations and take up very little space.

SPEAKERS AND HEADPHONES

Until a few years ago, good Hi Fi speakers came in medium-to-large wooden enclosures. This seemed necessary to obtain acceptable low-frequency response. In recent times, however, speakers have been developed which give reasonably good results in much smaller enclosures. And some speakers have been designed to include two variable "acoustical controls" which make it possible for the owner to further adjust both mid-range and treble tones to suit individual listening tastes.

An important point must be emphasized here: Always encourage and allow prospective customers to compare a number of different speakers in your listening room. Let the customer select the speakers he likes best after the listening test. If you follow this rule in all cases, you'll seldom have kick-backs. Remember, however, if a prospect makes it clear in the beginning that he wants the best possible Hi Fi reproduction, set up only the best speakers for his comparison test.

Speaker cabling may be shielded or unshielded. But the run length should be kept short. Under certain circumstances, shielded cable may be desirable. When no danger of 60-Hz hum pickup exists, 4-wire flat pressure-sensitive adhesive cable may be used. This comes in various colors for matching walls and it works nicely on painted plaster walls, wood paneling, wall baseboards and under rugs. In fact, direct under-rug speaker cabling frequently allows shorter runs well away from house wiring.

Headphones, designed for both monophonic and stereo use are a desirable and profitable accessory to Hi Fi stereo equipment. They are used primarily when one member of the family wishes to listen while others study, read or sleep. Headphones come in various types, some having three conductor plugs as shown in Fig. 12-2 and others having terminal lugs as shown in Fig. 12-3. Adapters are normally used between the amplifier output and phones to avoid overloading the phones. Most Hi Fi amplifiers made today employ double-pole-double-throw (DPDT) switches to disable both speakers and connect the amplifier outputs to a phone jack.

GENERAL CONSIDERATIONS

Two other important technical considerations should be briefly mentioned here. And, believe it or not, at this late date, both are controversial. But then, there probably are some people still alive today who insist that the earth is flat.

The first consideration has to do with the required frequency response of good audio amplifiers. Some "experts" still maintain that 20 Hz to 20 kHz is adequate, since this is somewhat better than the average range of human hearing. In one sense, this is true. But there are other considerations. And what these "experts" seem to forget is that with music reproduction we are not basically concerned with the frequency response of human ears. We are primarily concerned with an amplifier designed to provide reproduction fidelity. Actually, a well designed audio amplifier (having a very low order of harmonic and IM distortion) has not yet been made without its bandwidth automatically expanding to at least 100 kHz or higher. When an amplifier is designed that has a flat frequency response confined to a bandwidth from 10 Hz to 20 kHz and has an acceptable percentage of harmonic and IM distortion (0.2% and 0.2% respectively at full rated power), then we'll sell it door-to-door. We have not seen such an amplifier yet. The only amplifiers which have a very low order of harmonic and IM distortion are those which have reasonably flat responses to and beyond 100 kHz. So, the question is not frequency response but fidelity capability. Hence, the argument is strictly semantic.

The other point has to do with amplifier power requirements. The confused "expert" begins by asking, "How much power do you need in an average large living room? One watt?" And he usually adds, "Certainly not much more." He is right. But he also forgets other necessary considerations. We will mention only one here. Make no bones about it, to obtain a modest degree of musical realism without distortion —assuming that no severe acoustical problems exist in the average large listening room and depending on the speakers' efficiency—two good amplifiers with a minimum continuous RMS output power of approximately 40 to 50w are required. The amplifiers will be running at about 0.5 to 1w normally, but musical peaks can momentarily drive them to maximum output capability—or even into overload! And if amplifiers distort on peaks, we must reduce the sound pressure level (SPL) somewhat below the point where "concert realism" is obtained. The discriminating listener will become unhappy.

Although the trend is toward integrated-component equipment, some prospective customers may want separate tuners, control-unit/preamps and amplifiers. In any event, keep all signal-carrying cables as short as you can and use shielded cable for connections from component to component, from record players, tape players or tape recorders. This may not be necessary with speaker leads, as previously mentioned.

A long-standing "gripe" of the music loving public, especially that section of it which buys package-type AM/FM radio and phono combos, has been poor FM stereo reception. In some cases this complaint has been caused by inferior equipment and faulty station transmissions. But, more often than not, the fault has been insufficient antenna gain or multi-path reflections caused by an improperly designed or oriented antenna system. A 300-ohm folded dipole antenna made from flat TV lead-in is seldom satisfactory for FM stereo reception, even if cut to the exact frequency of the transmitter, oriented properly, and with the receiver located only a few blocks from the transmitter. To complete a satisfactory home audio system, make sure the FM antenna installation is the best possible.

One final important consideration is a carefully designed program of customer education. Teach your customers, by whatever means, to care for their equipment. First, give careful instructions on how the various controls of the pre-amp, amplifier, or speakers should be used. And what is more important, how to handle a phonograph. Explain that delicate pickups and styli cost money and should be handled with care. Explain that phonograph records and pre-recorded tapes last much longer and provide better reproduction if properly cared for. Let your customers know that a tape recorder or player needs periodic preventive maintenance— especially cleaning and head degaussing. A properly planned and executed educational program will provide your business with valuable dividends now and in years to come.

CHAPTER 13

Commercial Audio Systems

Traditionally, the commercial audio communications equipment (CACE) area of the electronics field has been called "public address" (PA). But this definition is so narrow that it is thoroughly ambiguous since the CACE area now includes not only PA systems (fixed and mobile), but intercom and paging equipment, background music systems, teaching lab and various other audio-visual equipment systems employed in business, industrial and financial establishments, educational, religious and medical institutions.

PUBLIC ADDRESS EQUIPMENT

PA systems are needed in today's world for two basic reasons: (1) To "boost" the voice of a person or persons so all listeners in a specified area can hear distinctly or (2) distribute the voices over auxiliary transducers located in other areas beyond the main listening area. These auxiliary transducers may be located in one or more other enclosed areas or outside the building where an overflow audience has gathered. It should be made clear that a so-called PA system can be used also for paging or for music distribution if it is so designed. Likewise, a background music system can serve as a paging system if properly designed. And intercom systems employed in some public places, like restaurants, for example, also can be used effectively as both background music and as paging systems.

The essential components of commercial audio installations include amplifiers, various specialized speaker types, microphones, distribution transformers, proper cabling, connectors, and other necessary hardware.

PRELIMINARY CONSIDERATIONS

Commercial audio, like TV, is a specialized business and requires considerable specialized technical study. And before we attempt to sell, install, or lease our first audio communications system—and especially before making a survey or quoting prices—it is necessary to briefly review some important preliminary considerations.

An amplifier's quality is our first consideration. It must provide the necessary power to operate a certain number of speakers at the necessary loudness level. And we don't mean that the amplifier should operate at its maximum output power rating. Additionally, our experience indicates that it would be disastrous to recommend that you dabble in "economy-type" audio amplifiers, speakers, or other necessary audio components.

Check the manufacturer's power rating specifications carefully. The output power rating of an amplifier for commercial use should be based on continuous (RMS) sine-wave power. And remember, the power at the amplifier output usually falls off some at the high-frequency end, even more at the extreme low end. Find out what the total harmonic content (THC) is at the amplifier's maximum power rating. Check the frequency at which these ratings were made (generally at 1 kHz). Some audio system designers hold that a general use PA amplifier (not including music distribution) should not have more than 5% THC. And this is relatively high when compared to good Hi Fi amplifiers. Avoid an amplifier that has its power rating specified as "maximum usable."

The next outstanding consideration is amplifier frequency response. Here again, there are plenty of ambiguous "ratings." Avoid amplifiers that merely give a single rating like: "Response—20 Hz to 20 kHz." An amplifier should be rated in plus or minus so many db either across the full response spectrum or at a number of specified frequencies within the total spectrum. Inquire if the response rating is at the amplifier's full-rated output or at some lower power. The output transformers in amplifiers vary widely in quality and cost.

Another important consideration, especially when selecting

an amplifier for both PA and music reproduction—from turntable, tape player, or whatever—is input sensitivity. This is closely related to output power. The preamp or first audio stage must be properly driven to provide, in turn, sufficient undistorted driving power to obtain the required undistorted amplifier output. Make sure you can depend on the amplifier's sensitivity rating. If you have doubts about <u>anything</u>, remember that every dependable manufacturer is always glad to guarantee his equipment specifications. Ask him about it.

Amplifier stability is another important factor. As indicated previously, the dividing line between an amplifier and an oscillator is very thin—like that spider's thread which separates dusk from darkness. Unless properly designed, the amplifier may "take off" at any time when long speaker lines are used. An output transformer having very low leakage inductance and low distributed capacity should be specified. These are some important factors involved in PA system amplifier selection.

Other lesser but frequently important factors may include both low- and high-impedance microphone inputs, area speaker selector switches, and a variety of output impedance taps. It should be mentioned here, also, that an amplifier having a "constant-voltage" output should be selected if a flexible system is desired or one having a larger number of speakers is planned.

AVAILABLE AMPLIFIER TYPES

We are generally concerned with two amplifier types in commercial PA equipment (1) combination preamp/amplifiers constructed on one chassis in one housing and (2) straight amplifiers having a separate preamp which may also include "mixing" facilities. A separate preamp is most often used in more elaborate, higher-powered installations.

The combination preamp/amplifier units, by far the most widely used, are in the low- and medium-powered category. These are available in both solid-state and electron-tube types. The units are made in a variety of configurations to satisfy almost any application in power outputs ranging from 10 to 50w.

Separate amplifiers, without built-in preamps, range from about 30 to 250w and higher. Of course, for a given application, the output power needed is determined by the number of

Front view of a Bell PA modular amplifier (top).
Rear view of the same amplifier (below).

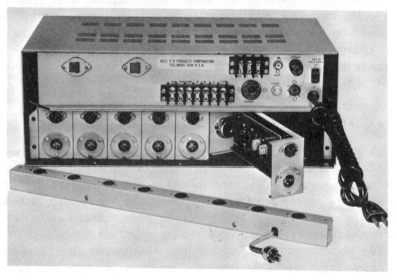

speakers in the system, the average wattage level at which all speakers are operated, plus acoustic and noise level conditions of the location involved. And it is always wise to allow a 20% or higher power overage. In some cases, a much larger power reserve may be considered to allow for future expansion—although additional separate amplifiers may be added if the expansion involves a rather large number of additional speakers.

Some audio equipment service-dealers specialize in custom-designed installations to satisfy the needs of particular situations. Some of these arrangements may contain a preamp/control or program mixing unit, one or more amplifiers, input equipment (tape player, turntable, AM/FM radio tuners), plus one or more microphones.

Various portable-type amplifiers and speakers, designed for fixed and mobile service, AC or battery operation, are readily available. These units are ruggedized, have carrying handles, and can easily be moved from place to place. The mobiles are generally adapted to 12v negative-ground operation. Some smaller portable units have shoulder straps for personal mobile use. And hand-held battery-powered integrated amplifier/speakers (or powered horns) are also manufactured.

DISTRIBUTION PRINCIPLES

Although it is beyond the scope of this book to provide theoretical details on acoustic principles as they apply to commercial audio distribution, the subject must be given a cursory look. Some acoustic principles we must be aware of are: (1) longer-than-acceptable reverberation times (2) initial reverberation level and (3) the techniques involved in solving these problems in audio distribution.

Depending on the material used in the room floor, walls, and ceiling construction, audio intelligence projected by a speaker will be reflected back and forth from these surfaces (some reflections being stronger than others), and into the listener's ears. This jumble of "echos" deteriorates the intelligence. We call this an "over-live" environment and in some cases, the "louder" we make the intelligence, the more unintelligible it becomes. Depending on the environment, reverberation times of more than a second (the length of time it takes an echo to die out), usually pose a distribution problem. But these problems can be solved satisfactorily in most cases.

We obviously cannot rebuild the already-constructed areas in which we are called upon to install effective audio systems. Hence, we must design satisfactory systems by selecting the most effective speaker types, the proper number, place each

Cone-type speaker designed especially for electronic musical instrument use. (Courtesy Jensen Mfg. Div., The Muter Co)

in the correct location, and supply each with the optimum audio power. In some cases this takes a bit of doing.

For example, it may be necessary to project the audio intelligence at a high-power level (where background noise is high but reverb is a minor problem) or locate the speaker or speakers closer to the listener and project the intelligence at a much lower power level. We may also find it necessary to run different speakers in a system at different power levels —usually dependent on how far they are located away from highly-reflecting surfaces. And where an environment has highly sound-absorbent material on floors, walls, and ceiling, we may have to provide a higher-powered amplifier to make up for the amount of power absorbed. We call this a "dead" location, which presents its own peculiar problems.

We must remember also, where a large number of listeners

are involved in an enclosed environment, considerable audio power is absorbed by the audience, and the acoustical characteristics change when the room is empty. When the room is half filled with people, the conditions change again.

Of course, in recent years a system has been developed by Dr. Charles Boner, a physicist, and his son, which goes a long way toward solving most reverb (echo) and acoustic gain problems. The Boner "equalization" system, in a simplified way, uses (1) broadband equalizers to obtain a "flat" response from the audio system; (2) zero-beats an oscillator with the acoustic feedback; (3) employs the oscillator output to tune a notch filter to the feedback frequency, and (4) places these filters in the transmission lines. A filter is employed for each feedback frequency arising in a particular location up to and including the full power output of the equipment. Thus far, the system is on a "custom-made" basis and is rather tedious, time-consuming, and beyond the practical application of the average installer or audio technician.

BASIC SPEAKER TYPES

As we now suspect, we cannot intelligently select the best amplifier for a given application without knowing something about the installation's environment. Likewise, we cannot select speakers for a given job unless we know some basic facts about various types. But here, in the long run, you

Heavy-duty driver unit. (Courtesy University Sound)

must lean heavily on information which can be obtained from reputable speaker manufacturers.

You may already know that speakers are designed in two basic forms—state-of-the-art developments determining the number of varieties of these two forms at any point in time. Both speaker forms employ moving voice coils. And the traditional form has a large-area fiber cone. This speaker is normally mounted on a wooden baffle and placed in an enclosure. The other basic design, sometimes called a "compression driver" or "pressure driver," has a voice coil and a small-area diaphragm enclosed in a housing which is threaded at its front to accept the small end of a "horn" or "trumpet" structure. These structures are made in various shapes, lengths, and diameters for specific results. We will use a variety of both speaker types in our installations and perhaps the pressure drivers will predominate for a time in large-area audio distribution —especially in those areas where high noise levels and unique reverb characteristics prevail. But this "perhaps" viewpoint may not necessarily hold true tomorrow. Research and development in the "large-cone" speaker area does not stand still.

SPEAKER CHARACTERISTICS

Driver-type speakers having responses from approximately 70 Hz to 10 kHz are available. Cone-type combinations have a much wider response. The driver types are easy to weather-proof (even waterproof) and like comparable cone types, come with or without line - matching transformers. Drivers have various wattage ratings at full-range and at so-called "adjusted (limited frequency response) range."

Remembering our qualifying remarks about cone speakers, driver-type speakers employing the "reflex trumpet" principle provide much higher "acoustical conversion efficiencies" when comparable driving powers are used. This may or may not be an advantage, depending on the particular application with which you are concerned. A directional reflex trumpet-type speaker is shown in Fig. 13-1. According to the manufacturer's specifications it has certain characteristics which make it especially adaptable to certain applications. It is not entirely clear at first glance what characteristics are for what application but further inquiry and study will probably make it somewhat clearer. You are encouraged to press the manufacturer for further information. Or, better still, invest a few dollars and take that speaker into a number of locations, check it out, and find out for yourself what it will do under various circumstances. For this, a noise-level meter may come in handy.

Reflex trumpets are made in various lengths, having different "dispersion angles." They come in "narrow" and "wide-angle shapes. And, of course, you can gang or "stack" them like TV antennas both vertically and horizontally to achieve certain effects (see Fig. 13-2). And the results you obtain depend entirely on the specific environment where they are used.

Some classical examples of the dispersion angles, coverage, and penetration characteristics of the two types may give you some ideas. If an area to be covered with audio intelligence is much wider than it is deep, a wide-angle horn having a 120° horizontal and 60° vertical projection pattern might cover the area satisfactorily when mounted normally as shown in Fig. 13-3. But suppose the area is very narrow and deep. Here we may be able to use the same type horn if we turn it 90° as shown in Fig. 13-4. This would seem satisfactory only if the area depth is not greater than the penetrating ability of the trumpet.

If we mount a "directional" (round) horn in a certain area we come up with patterns as shown in Fig. 13-5A and B. This particular horn is designed to have a moderate dispersion angle which resembles a teardrop. Of course, the vertical

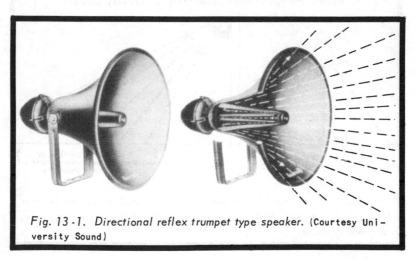

Fig. 13-1. Directional reflex trumpet type speaker. (Courtesy University Sound)

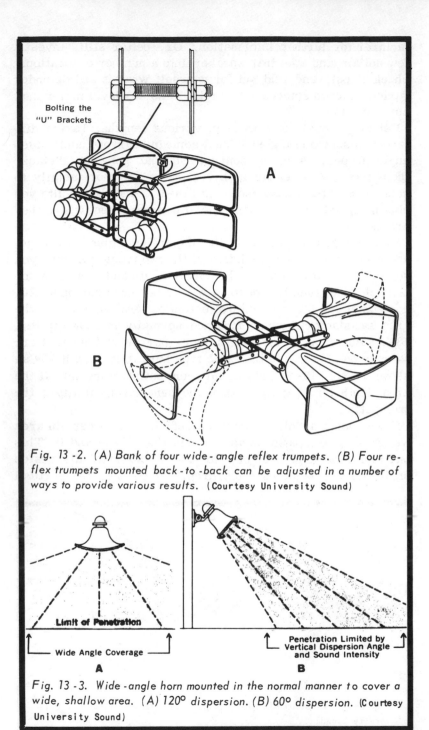

Bolting the "U" Brackets

A

B

Fig. 13-2. (A) Bank of four wide-angle reflex trumpets. (B) Four reflex trumpets mounted back-to-back can be adjusted in a number of ways to provide various results. (Courtesy University Sound)

Limit of Penetration

Wide Angle Coverage

A

Penetration Limited by Vertical Dispersion Angle and Sound Intensity

B

Fig. 13-3. Wide-angle horn mounted in the normal manner to cover a wide, shallow area. (A) 120° dispersion. (B) 60° dispersion. (Courtesy University Sound)

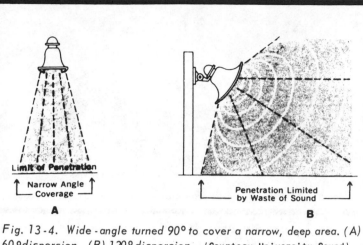

A

B

Fig. 13-4. Wide-angle turned 90° to cover a narrow, deep area. (A) 60° dispersion. (B) 120° dispersion. (Courtesy University Sound)

Fig. 13-5. These sketches show the representative areas covered by a round horn having a moderate dispersion angle. (Courtesy University Sound)

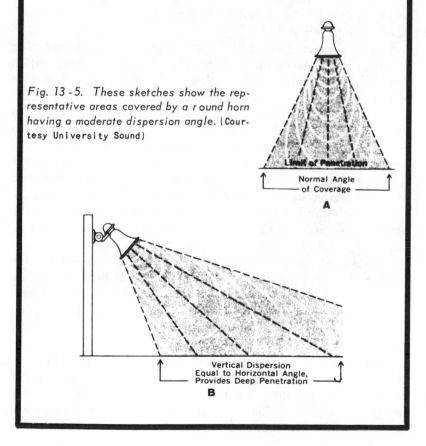

A

B

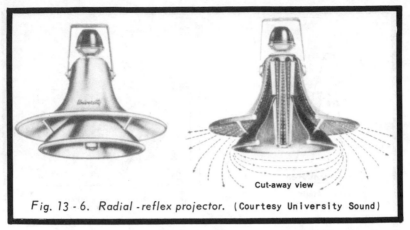

Fig. 13 - 6. Radial -reflex projector. (Courtesy University Sound)

dispersion and audio energy is the same as it is in the horizontal plane—giving considerable penetration. The same pattern as shown in Fig. 13 - 3 can usually be duplicated by the round horn when it is mounted at the proper angle to obtain maximum penetration.

A group of driver - unit horns are designed to be mounted in ceilings above the heads of listeners. These designs, called "radial reflex projectors" or "radial horns," are shaped as shown in Fig. 13-6. They are designed to provide full 360° uniform horizontal dispersion. These are available in a few sizes and are normally used in high-ceiling, large-area places like factories, gymnasiums, warehouses, etc. Here, the ceiling height, power used, or the number of speakers in the installation are prime considerations. The speaker is useful in areas having only moderate background noise levels. Charts are available from the manufacturer which give an idea of the square feet and circular - area coverage, mounting heights, etc., for various speakers in this category.

Low-ceiling, flush mounting type speakers are available for paging and background music distribution. In fact, a speaker is available for every conceivable type of normal and unusual application, including those designed for underwater use (Fig. 13-7), and dual - channel access types having double voice coils as shown in Fig. 13-8.

Another speaker which has specialized characteristics, is the columnar, or column, sometimes called a "line radiator." It is made up of a long array of small woofers and a shorter array of tweeters. The "effective length" of the arrays are made to vary with frequency by employing signal distribution shaping networks for each element as shown in Fig. 13-9. A number of columnar speakers are frequently spaced in an area to give overall audience coverage without too much energy bouncing back and forth from floor and ceiling. They can be used in various arrangements to provide satisfactory reinforcement, minimize feedback and other acoustic problems

LINE MATCHING

Depending on a variety of factors, especially the line lengths from amplifiers to speakers, we have to install proper lines and use various methods of matching lines to the amplifier and the speakers to the lines.

Some simple, direct arrangements for small systems having short lines are shown in Fig. 13-10. In the aforementioned

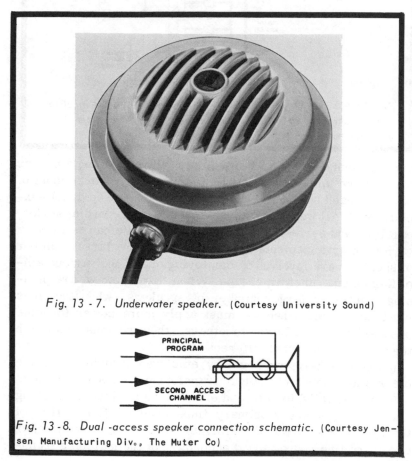

Fig. 13-7. Underwater speaker. (Courtesy University Sound)

Fig. 13-8. Dual-access speaker connection schematic. (Courtesy Jensen Manufacturing Div., The Muter Co)

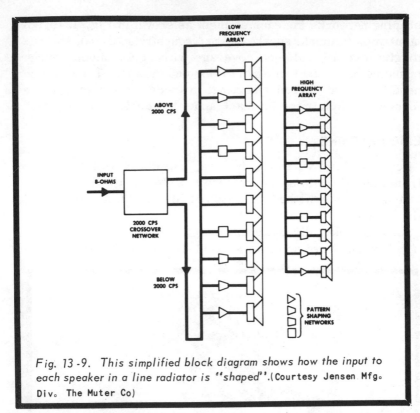

Fig. 13-9. This simplified block diagram shows how the input to each speaker in a line radiator is "shaped".(Courtesy Jensen Mfg. Div. The Muter Co)

arrangements we're concerned only with providing a proper load for the amplifier and with feeding the proper amount of power to each speaker. These arrangements seem self-explanatory. All we need to know is Ohm's law regarding series, parallel, and series/parallel impedances. We can work out a variety of arrangements where five, six, or a larger number of speakers are involved by employing one of the various amplifier output transformer taps. When the system we are planning requires a larger number of speakers, when the lines must be longer, when we must apply more power to some speakers and less power to others, then we must approach the problem somewhat differently.

In a general way, we begin by employing a high-impedance tap on the amplifier's output transformer secondary (125 to 600 ohms usually) for one end of the line and place a matching transformer having a primary impedance similar to that of the amplifier's output transformer secondary at the speaker end of the line. The tapped secondary of the matching transformer is then used to match a speaker or combination of speakers. A schematic of a typical matching transformer is shown in Fig. 13-11. The schematic of a long, high-impedance line which uses a matching transformer is shown in Fig. 13-12A. The transformer primary matches the high-impedance line and its secondary matches the 16-ohm speaker.

If two 16-ohm speakers are close together they can be connected to the output of one matching transformer as shown in Fig. 13-12B. If they are spaced a considerable distance apart, two separate matching transformers must be used as shown in Fig. 13-12C. Four 8-ohm speakers can be connected series/

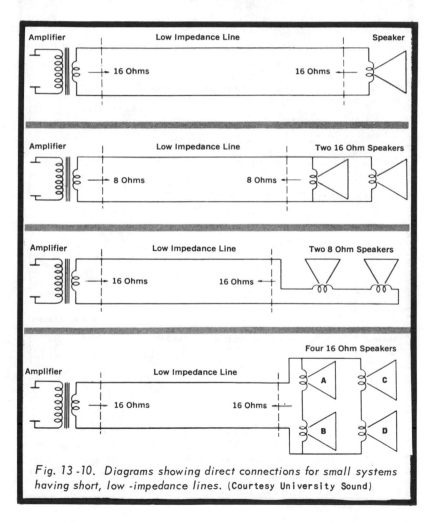

Fig. 13-10. Diagrams showing direct connections for small systems having short, low-impedance lines. (Courtesy University Sound)

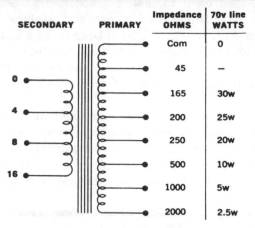

		Impedance OHMS	70v line WATTS
		Com	0
		45	–
		165	30w
		200	25w
		250	20w
		500	10w
		1000	5w
		2000	2.5w

SECONDARY   PRIMARY

Fig. 13-11. Schematic of a low-to-medium impedance matching transformer.

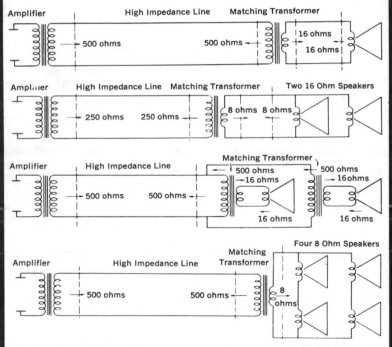

Fig. 13-12. (A) Long, high-impedance line using a matching transformer for one speaker. (B) Two 16-ohm speakers on one matching transformer. (C) Two matching transformers for two speakers spaced a considerable distance apart. (D) Four 8-ohm speakers in series/parallel on a long, high-impedance line. (Courtesy University Sound)

parallel across one matching transformer if the transformer is designed to handle the total load, as shown in Fig. 13-12D.

Suppose we need to install a paging/talkback system in an area, for example, where the noise levels vary widely—perhaps in a small factory. One part, the manufacturing area, requires two 10w speakers. Another part, less noisy, requires two 5w speakers. Still another part, perhaps an enclosed stock room with various supply depots plus an office area, requires a total of 20 one-watt speakers. At a glance we see the total amplifier power required adds up to 50w. Because of the relatively long transmission lines, plus the difference in power requirements, we will need separate matching transformers for each speaker.

Additionally, while the speaker runs are long and a high-impedance line is needed, we must consider the availability of matching transformers having the proper primary impedances in a parallel setup like this. With this consideration in mind, we'll select a 250-ohm tap on the amplifier's output transformer.

Next, we have to figure how to get the required power to each speaker. There are three speaker groups. We need a ratio between the total power and the power required for each speaker in each group. We divide 50w by 10w and get a 5:1 ratio. Then 50w is divided by 5w to obtain a 10:1 ratio. And finally, 50w is divided by 1w which gives a 50:1 ratio. Now we multiply the line impedance (250 ohms) by 5 which gives 1.2K—the impedance of the primary of each matching transformer to be used with the 10w speakers. We do the same for the 5 and 1w speakers and come up with 2.5K and 12.5K, respectively. If we use the reciprocal formula for parallel impedances (1 divided by 1/R1, 1/R2, 1/R3, etc.) we find that all transformer impedances here come out to 250 ohms—the line impedance. Of course, the matching transformer secondaries must also match the voice coil impedances of the speakers.

CONSTANT-VOLTAGE LINES

As previously mentioned, when we use the constant-voltage system, the amplifier must have an output transformer with a tap which delivers 70.7v (or 25v, and in amplifiers having an output of over 100w, a 141v tap) at the amplifier's full-rated output. This system allows additional speakers to be added to existing systems without having to make extensive impedance calculations and also makes it easy to arrange various amounts of drive to different speakers in an audio system. It must be understood, however, that the total speaker power must not exceed the amplifier's maximum power output. If the total amount of power used by all speakers is equal to or less than the amplifier's rated power, the system will be properly loaded.

Let's not misunderstand the system we are working with here. A "70.7v speaker distribution line" does not mean that the voltage on the line will necessarily be a constant 70.7v—unless we are running the amplifier at full-rated power. And even then, it will vary—depending on the variations at the amplifier input. It would be constant only if a steady sine-wave signal is injected into the amplifier input. With any amplifier of less than 100w, the maximum RMS voltage of a sine wave at the amplifier's full-rated output would be 70.7v. Normally, matching transformer primaries will have the same amount of voltage. This is what is meant by a "constant-voltage" system.

Constant-voltage transformers are arranged so the primary taps supply various amounts of power and secondary taps provide a match for various speaker impedances. The primary taps may be marked in watts, in ohms, or both. If marked in watts, calculations are simplified. But the mathematics is simple even if the taps are marked in ohms. You can use Ohm's formula ($Z = E^2/P$) to find the impedance tap required. In the case of a 70.7v system, if you want to feed 5w to a speaker, square 70.7 and divide by 5 and you get an impedance of 1K. Actually, on a 70.7v line, you need only divide 5000 by the speaker power in all cases, and you'll come up with the proper impedance. A nomograph (or nomogram) also may be used to tell you almost instantly which impedance tap must be selected on the speaker transformer primary to provide the desired power on a certain speaker.

If we use the previously described paging/talkback small factory audio system on a 70.7v distribution system, we find the necessary matching transformer primary impedances as previously described: $Z = E^2/P = 5000$ divided by 10w (speaker) power = 500 ohms. Impedances for the 5 and 1w speakers are derived in the same way (which come to 1K and 5K, respectively).

Among other important points that must be considered is wire size and audio transmission cable lengths. For example, maximum length of a 4-ohm cable made from two size 20 B&S gauge wires should not exceed 25'. An 8-ohm line using 20-gauge wire can run for 50' and a 16-ohm line to 100'. A 500-ohm 20-gauge line can run up to 1500' without excessive loss. Lines made from size 18-, 16-, and 14-gauge wire can be made progressively longer. A size 14-gauge, 500-ohm pair, for example, can be run up to 5000'. It should also be mentioned that volume levels on individual speakers can be varied by installing T or L pads.

MICROPHONES

The equipment we have previously discussed in this Chapter is all in the "high-level" distribution category. Microphones and their cables are in the "low-level" area. For this reason, our problems with microphones are somewhat different. It must be made clear once again that it does not pay to select any but the best microphones, anymore than it makes sense to select economy-type preamps, amplifiers, speakers, or other components used in a commercial audio system.

We are not too much concerned here with the differences existing between the various microphone types—carbon, dynamic, crystal, ribbon, ceramic, and magnetic. This does not mean that the characteristics of a certain type microphone may not become a prime consideration in some particular application. But in an overall, general sense, we are much more concerned with frequency response, directivity, and to a lesser extent with sensitivity and impedance. Most often, however, ribbon or dynamic mikes are used where both voice and music applications are necessary.

Although it is sometimes said that a "good" general purpose commercial microphone should have a flat frequency response from 40 Hz to 15 kHz, plus or minus 2 db, few microphones meet these specifications. Since PA-type microphones are generally used for voice only (music is fed directly to a preamp input), we feel that a microphone having a response from 70 Hz to 10 kHz, plus or minus 2 db, is thoroughly adequate for PA-type applications. Indeed, for straight PA and indoor

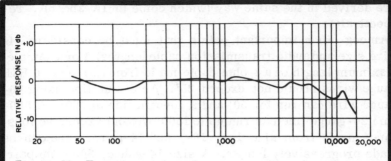

Fig. 13-13. Typical microphone frequency response chart. (Courtesy Shure)

paging systems, microphones having a relatively flat response from 200 Hz to 5 kHz are frequently employed with good results. Of course, if you are installing microphones for a church choir or a stage where "live" music is to be reinforced, this is an entirely different matter. If this is the kind of system you plan to install, by all means make sure that the microphones have the widest and flattest possible frequency response. And always insist on a frequency response chart (see Fig. 13-13) of the microphone or microphones you plan to use in a system before purchasing.

It should be understood that we are not interested in "flatness" of a microphone's frequency response just for the sake of frequency response flatness. We are concerned primarily with sensitive peaks in the microphone's response which often aid in causing undesirable feedback at relatively low output levels. In a general sense, a microphone which has a plus or minus 2 db response from 70 Hz to 10 kHz is more desirable than one which has an "average" response of 40 Hz to 20 kHz but having one or more 3 db or higher peaks in its low- or medium-frequency response range.

All this does not mean to indicate that we must not, under certain circumstances, select microphones which have been deliberately "imagineered" to fit special applications. In this case, the response curve may not conform to generally desirable characteristics. In fact, we may even "dope-up" a microphone input with an equalization network as one approach to minimizing feedback.

Our second prime consideration in selecting microphones is directivity, or pick-up pattern. We have three directivity patterns to consider as shown in Fig. 13-14—indicating unidirectional, omnidirectional, and bidirectional microphone characteristics. We will seldom use the latter two types except in special cases.

The unidirectional pattern is that of a cardioid (heart-shaped response) microphone. The typical cardioid generally provides 20 db or more reduction in pickup directly from its rear. It reduces pickup 6 db or better at 90° each side of its front. The directional or polar response curve of a cardioid microphone is shown in Fig. 13-15. Cardioids are designed to have various angles of sensitivity. The typical cardioid is sensitive over a range of about 150°. Others have a range of about 80°. The so-called "ultra-cardioids" have sharp frontal response lobes with a sensitivity range of about 40°. In many applications, cardioids serve as excellent microphones where speakers are mounted in the back and sides of a room or hall and thus minimize feedback. Audience noises are also suppressed.

Omnidirectional microphones are generally hand-held and little used except in special areas of commercial audio where feedback and audience noises do not cause problems. Bi-directional microphones are generally in the same category, except they are ideal for interviews or when two groups are working opposite each other or side-by-side with the "figure-eight" microphone placed between them. Some other microphones which we may find useful under certain circumstances are controlled-magnetic, hand-held and controlled-reluctance cartridge type hand-held noise-cancelling microphones. These are excellent for mobile PA work. The noise-cancelling type

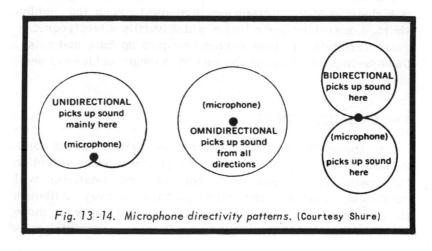

Fig. 13-14. Microphone directivity patterns. (Courtesy Shure)

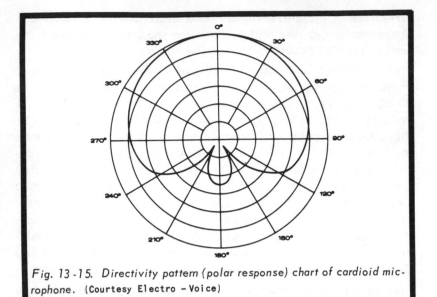

Fig. 13-15. Directivity pattern (polar response) chart of cardioid microphone. (Courtesy Electro - Voice)

is particularly useful in applications where high background noises prevail.

As previously mentioned, the "sensitivity" of a microphone is one of the lesser importances that concern us. This is especially true because, as in most other sensitivity measurements, microphone sensitivity has been equated with "gain." And since we depend primarily on preamplifiers for necessary gain, this makes microphone "sensitivity" of little importance in PA and other commercial audio work. Remember, though, if high-impedance microphones are used, keep the cables short. Long cables cause losses which usually affect frequency characteristics and long cables can pick up hum and noise. Use low-impedance microphones where longer cables are necessary.

FINAL CONSIDERATIONS

It is impossible here to cover all the "tricks" of the commercial audio area. But the most helpful component selection information you will ever obtain for any given installation will come when you make the initial technical survey. Although the prospective customer will usually present you with more ideas than your experience can immediately convert into practical application, you can rest assured that he will cue you in on the important points to be considered, as he explains what he wants the installation to do. And a careful survey of the installation site should be the first step in a successful audio installation.

The most experienced installers rely only on actual tests made in the area to be covered by an audio system. A test setup need not be elaborate. A medium-powered amplifier having a 70.7v output and one or two matching transformers (primarily so the input power to the speakers can be varied and precisely checked), plus a microphone or two and perhaps a tape recorder or player is about all that is required. A noise-level meter is helpful in high noise-level areas. An inexpensive slide-rule type calculator provided by Jensen can also prove helpful.

Each job, large or small, must be approached in a well-planned, systematic manner. And the survey includes not only the job of obtaining information which makes it possible for you to intelligently select major components, but an opportunity to measure and sketch floors, walls, ceilings, and possibly loft or attic areas; determine the length of cable runs, the amount, size and type of cable required; prepare a detailed list of all components and accessories and finally, figure the cost estimate—making certain that sufficient labor for a reasonable profit is included.

CHAPTER 14

Home and Business Intercoms

Although intercom (intercommunications) systems have been used by business and industrial establishments for years, home intercoms were little more than novelties less than ten years ago. Since then, many improvements have been made in both home and business type intercoms and more intercoms are being sold to home owners and apartment dwellers. Many are being installed in new homes during construction.

Home-type intercoms are now available in various types ranging from simple transistorized two-station "baby-sitters" to multi-room types having 10 or more stations, plus AM or AM/FM radios. Some types have individual master switches to control all room stations, talk/listen and radio volume controls at substations. Some are designed so the intercom volume level overrides music volume or automatically switches off music when the intercom is used. Others have phone/tape inputs and various other features. Outdoor type speaker/microphones make it possible to answer the front door from the kitchen or from any room in the house or monitor a baby's room or nursery from any room, including the kitchen, workshop or garage. Even the jargon used to describe the two primary intercom units has changed. What was once called the "master" station is now frequently called the "staff" station. "Slave" stations are now called "remotes," or substations."

Intercoms can be arranged in a variety of ways. For example, all stations can be arranged as "masters"; a remote station may be arranged to transmit but not initiate a call (used as "baby-sitter" monitor); or a remote can initiate a call to one master station. In fact, almost any number of arrangements are possible.

Few differences exist between the average home and business-type intercoms—except business intercoms normally do not have built-in radios and handsets are frequently used on business intercoms for privacy—the speaker is automatically disconnected when the handset is lifted from its cradle. Business intercoms are usually desk mounted and home intercoms may be desk, table-top, shelf, or in-wall flush mounted. And some large business intercoms have many more sub-stations than home-types do.

TYPICAL INTERCOM CIRCUITS

Of all the electronic equipment employed in audio communications, the every-day home or business intercom is the simplest. Basically, an intercom system consists of an amplifier and two speakers—one speaker located in another room—both speakers being used as microphones while talking. In practice, however, there is a little more to them.

Two basically different types of home and business intercoms are normally made: Wired and "wireless," or RF type. Cables containing a number of wires, the number dependent primarily on how many sub-stations are employed, are normally used to connect the various stations in the wired type. The RF type, employing the "carrier current" (or carrier frequency) method, uses the AC power line instead of cables. The unit is simply plugged into an AC outlet. But this system will work only between points which have AC outlets common to a single power company step-down transformer. Ordinary RF systems are designed to operate on low frequency, which may be in the vicinity of 100 to 300 kHz. Some systems have switches which provide for both cable and house AC wiring operation. RF systems are usually supplied in pairs, although additional units can be added to other rooms.

WIRED SYSTEM

The schematic of a small 6-station cable-type intercom system is shown in Fig. 14-1. This unit operates from a 9v battery but a power supply can be substituted. Most home and business intercoms made today have solid-state power

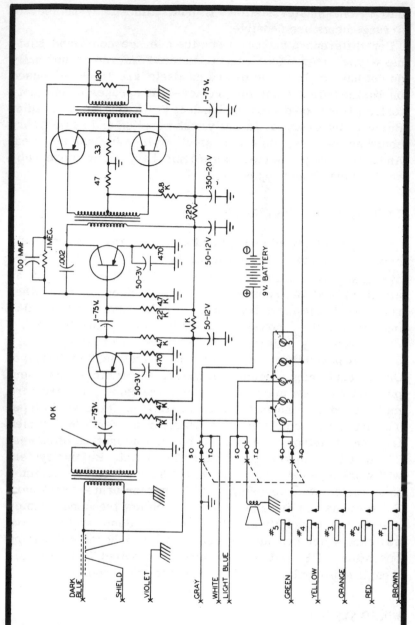

Fig. 14-1. Schematic of a 6-station, transistorized cable-type intercom. (Courtesy Fanon)

supplies that employ a step-down transformer and operate from 117v outlets.

Three different types of remote stations can be used with this unit, as shown in Fig. 14-2. The simple "baby-sitter" monitor is shown at "A". It is always on and cannot be used to initiate a call for two-way conversation. The remote at "B" can call to one master and carry on a two-way conversation. The one at "C" is designed to call any one of up to six master stations and carry on two-way conversations.

It should be pointed out here that there are more elaborate commercial-type intercoms which employ a large number of stations but do not use multi-conductor cables. These systems employ a high-frequency carrier and subcarrier system which may be AM, FM, or SSB (single-sideband). A coax cable is used for all channels and bandpass filters select the various frequencies. Many of these systems are custom designed to fit a particular need. For example, it is said that a SSB system at the NASA (National Aeronautics and Space Administration's) operations and checkout building at Merritt

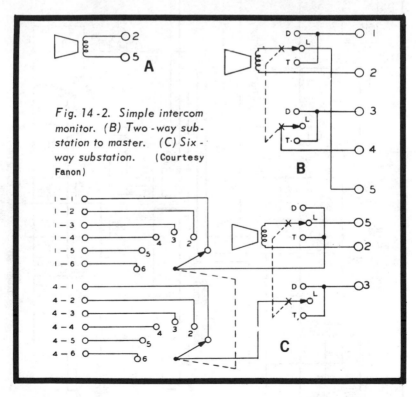

Fig. 14-2. Simple intercom monitor. (B) Two-way substation to master. (C) Six-way substation. (Courtesy Fanon)

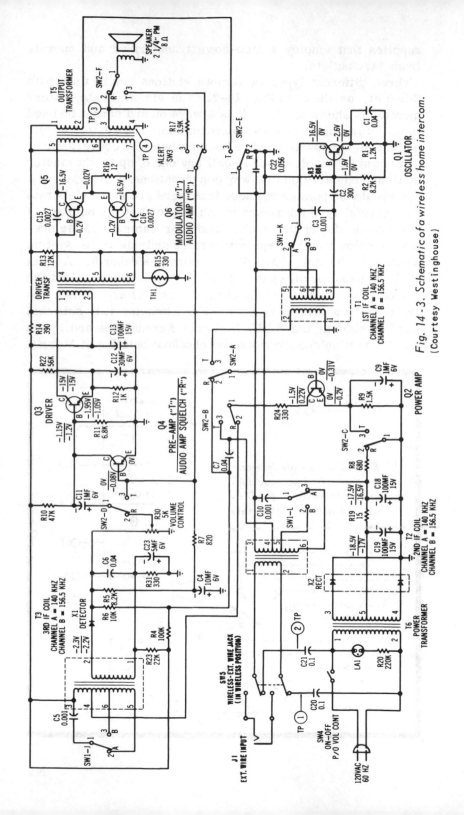

Fig. 14-3. Schematic of a wireless home intercom.
(Courtesy Westinghouse)

Island, Florida, uses a system which employs more than 100 duplex channels. This system is designed to serve offices and operational areas within the building.

RF SYSTEM

The schematic of an RF intercom is shown in Fig. 14-3. This particular system, designed for home use, may be operated on either of two frequencies—140 or 156.5 kHz. The frequency that gives the best results is normally recommended. Of course, all units in a system must use the same frequency for two-way communications. Typical of the RF intercom, this one employs basic transceiver principles. The 2-stage audio output serves for both receiving and transmitting (as an audio modulator for transmitting); the audio amplifier/ squelch serves as a "microphone" preamp. Q2 serves as an RF amplifier. In addition, Q1 is the transmitting oscillator and X1 is the receiving detector.

Do not be confused by the designations 1st, 2nd and 3rd "IF" indicated on the schematic. Since no heterodyning or frequency conversion takes place in the circuit—the oscillator tank (T1) and coils T2 and T3 are tapped to operate on either 140 or 156.5 kHz—no intermediate frequencies are employed in the intercom. Although not indicated on the schematic, the cores of T1, T2, and T3 are adjustable over a narrow frequency range both sides of the two frequencies employed.

The frequencies used on this intercom are not critical but it may be necessary to check and adjust the oscillator's frequency-determining coil, T1, and then align T2 and T3 to this frequency. T1 and T2 are adjusted while SW2 is in the "talk" position and coil T3 is adjusted while SW2 is released or in the "receive" position.

The oscillator frequency is checked with an accurate signal generator together with a scope. The test setup is shown in Fig. 14-4. Test and alignment points, as shown on the schematic, can be reached by removing the intercom's back cover. The simple steps for frequency check and alignment are as follows:

(1) Switch the RF signal generator, scope and intercom on and allow them to heat for 15 minutes.

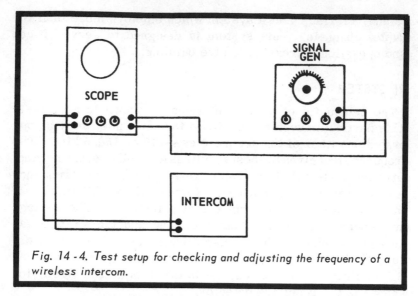

Fig. 14 -4. Test setup for checking and adjusting the frequency of a wireless intercom.

(2) Set the generator dial to 140 (or 156.5) kHz. Its output level should be at least 500 mw.

(3) Connect the generator's output leads to the scope's horizontal input and the scope's vertical input leads to the intercom's test points 1 and 2.

(4) Set SW5 on "wireless" switch SW1 to channel "A" (or "B"). Depress SW2 and the "lock-in" button to hold SW2 in the "talk" position.

(5) Adjust T1's core to obtain a Lissajous pattern on the scope screen like that shown in Fig. 14-5. This pattern indicates that the oscillator is operating at 140 (or 156.5) kHz —assuming the signal generator frequency is correct.

(6) Adjust T2's core for maximum amplitude (on the scope pattern). Then readjust T1 and T2 for greatest amplitude.

(7) Adjust T2's core 1/10 turn clockwise (CW) to insure stability.

(8) Now readjust T1's core to again obtain a Lissajous pattern on the scope screen as in Step 5. These two adjustments are necessary to stabilize oscillator operation.

(9) Disconnect the signal generator leads from the scope, the scope leads from the intercom, and use the VTVM and signal generator setup shown in Fig. 14-6. The signal generator output leads are connected to the intercom test points 1 and 2 and the VTVM leads to the intercom test points 3 and

4 (audio output). Modulate the generator's RF signal 30% with a 1-kHz audio sine wave.

(10) Release the "lock-in" button so SW2 will be in the "receive" position.

(11) Adjust T3's core for maximum deflection on the VTVM.

Notice that you can use either 140 or 156.5 kHz for the aforementioned checks and alignments but one or the other frequency must be used throughout the alignment procedure for all units.

OTHER PROBLEMS

Although you can make money by selling, installing, and

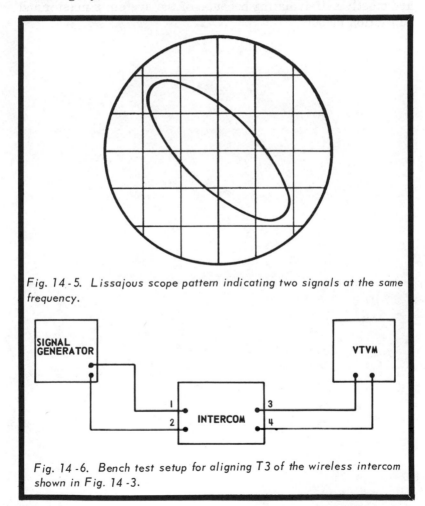

Fig. 14-5. Lissajous scope pattern indicating two signals at the same frequency.

Fig. 14-6. Bench test setup for aligning T3 of the wireless intercom shown in Fig. 14-3.

servicing home and business intercoms, one substantial section of the business is not easy for TV-radio service-dealers and technicians to obtain. This is new construction sales and installations, both home and business. The problem here is the conflict existing between the IBEW (International Brotherhood of Electrical Workers) and TV-radio workers, due to the fact that few TV-radio service-dealers or technicians have availed themselves of electricians' licenses or fail to employ one or more licensed electricians.

The technical problems are few, since manufacturers furnish installation details—including cabling diagrams. Maintenance is relatively simple because problems in the system are mostly self-isolating because of the system's master and substation layout.

Index

A

AC voltmeter, 13
"Acoustic" speaker control, 215
Advertising, 210
AFC, 48
Alignment, FM detector, 59
  multiplex, 70
  wireless intercom, 247
AM/FM auto radio, 169
Amplifier controls, 108
Amplifier troubleshooting, 110
Amplifier types, PA, 221
Amplitude modulation, 44
AM signals, 44
Analyzer, wave, 14
Antenna coil, 27, 34
Antenna system, FM, 218
Antenna trimmer adjustment, 167
Anti-skating, tone arm, 199
Attenuators, 87
Attenuation due to head mis-alignment, 149
Audio distortion, 34
Audio level meter, 18
Audio signal injection, 21
Audio transformer, 34
Automatic frequency control, 48
Automatic frequency response compensation, 88
Automatic tuning system, 173
Automatic volume control, 33
Auto radio, 167
Auto radio removal, 169
AVC, 34
Azimuth adjustment tape head, 153

B

Background music, 62
Background noise, 35
Balance control, 106
Balanced-bridge stereo detector, 79
Balanced multiplex detector, 67
Balance, output transistor, 35
Balance, tone arm, 202
Bass control, 88
Battery checks, 43
Bearing, tone arm, 199
Bench setup, auto radio, 169
Bias, Class A amplifier, 83
Bias oscillator, 141
Bias-reject circuits, 144
Bias stabilization, 82
Bidirectional microphones, 239
Boner equalization system, 225
Bootstrap capacitor, 85
Bootstrapped preamp, 85
Bootstrapping, 97
Brake static, 179
Bridge rectifier circuit, 128
Buzz test, 24

C

Cam, record changer, 191
Capacitance-multiplier, 136
Capacitive divider, 136
Capacitor leakage, 32
Capacitor noise, 35
Capacitor overload test, 40
Cardioid microphones, 239
Carrier current intercom, 243
Cartridge, tone arm, 202
Cassette recorder, 161
Cathode-heater leakage, 36
Change function, record player, 195
Channel separation, 73
Characteristic, dynamic transfer, 83
Class AB amplifier, 107
Class A driver, 83

Class B amplifier, 100
Cleaning, mobile tape player
  head, 185
  record changer, 207
  tape heads, 150
  tape recorder, 147
Click test, 24
  auto radio, 178
Cold intermittents, 37
Cold solder joint, 35
Column speaker, 230
Combination mobile radio and
  tape player, 183
Combination R/P, 141
Combinations, 118
Commercial audio, 220
Compatible FM stereo signal,
  61
Complementary-symmetry
  output, 101
Component replacement FM
  receiver, 55
Composite stereo signal, 71
Cone-type PA speakers, 226
Constant-voltage speaker line,
  235
Constant-voltage transformer,
  236
Converter, 28
  FM, 52
Converter checks, 25, 30
Coolant spray, 39
Crossover distortion, 107
Crosstalk, 73
Current balance, output tran-
  sistors, 35
Current consumption, 25
Current consumption measure-
  ment, 43
Current feedback, 96
Current measurement, output
  transistors, 35
  tape, head, 155
Customer relations, 210
Cycling, record changer, 191

D

Darlington output, 101
db meter scale, 14
db separation, 74
"Dead" environment, 224
Dead radio, 20
Dead spots, FM receiver, 55
Decoding, multiplex, 62

De-emphasis network, stereo
  tuner, 67
Degaussing, tape head, 151
Degenerative feedback, 91, 84
Delay lines, 182
Demagnetizing tape heads, 151
Demodulator, multiplex, 62
Detector, FM, 45
  stereo, 67
Diode stabilizing circuit, 96
Direct-drive output, 101
Directional reflex trumpet,
  227
Directivity patterns, micro-
  phone, 239
Discriminator, 47
Discriminator alignment, 59
Distortion, 14, 32, 34, 59,
  84, 107, 217
  FM, 55
Drift, 17
  FM tuner, 53
Drive, record changer, 189
Driver stage, 83
Driver-type PA speakers, 226
Dual-channel access speaker,
  230
Dual-source transducer, 61
Dynamic transfer character-
  istic curve, 83

E

Earphone jack, 34
Earphones, 216
Echos, 223
Electrolytic capacitors, 33
Electromechanical delay line,
  182
Emitter-follower, 84
Emitter-follower regulator,
  132
Emitter swamping, 96
Endless loop cartridge, 160
Erase head, 140
Erase oscillator, 141
Equalizers, 87

F

Feedback, 91
Ferrite loop, 27, 34
"Figure-eight" microphones,
  239
Filter capacitors, 36
Flutter meter, 17

Flutter, tape recorder, 145
FM antenna system, 217
FM converter, 52
FM detector, 45
FM IF, 45
FM multiplex, 61
FM multiplex transmission, 63
FM oscillator, 52
FM "quieting," 76
FM receiver sensitivity, 76
FM signals, 44
FM tuner sensitivity, 57
Foster-Seeley discriminator, 47
Frequency, bias oscillator, 141
Frequency modulation, 44
Frequency response, amplifier, 216
microphones, 237
tape recorder, 145
Frequency response compensation, 87
Frequency selective negative feedback, 92
Frying noise, 35
Fullwave rectifier, 127
Function switching, 120
Fundamental-suppression analyzer, 16

G

Generator, stereo, 75
Generator whine, 179
Grid-dip meter, 29
Grounded-collector amplifier, 84
Grounded-emitter amplifier, 82

H

Halfwave rectifiers, 125
Harmonic distortion measurement, 14
"Hash," auto radio, 179
Headphones, 216
Height adjustment, tape head, 153
High-frequency feedback, 95
High-impedance amplifier input, 221
High-impedance speaker line, 233

High-pitched squeals, 37
Hiss, FM, 55
Home "listening" room, 212
Horizontal movement, tone arm, 202
Hot intermittents, 37
Hum, 36, 111
Hum control, 112, 131

I

IC FM IF, 81
ICs, 115
IF, FM, 45
IF signal injection, 21
IF transformer, 27
IF transformer leakage, 35
IM distortion, 14
Impedance-matching stage, 84
Inadequate limiting, 57
Indexing switch, record changer, 202
Indicator, stereo, 69
Inputs, PA amplifier, 221
Intercoms, 242
Integrated circuits, 115
Interference, 112
auto radio, 179
Intermittent antenna, 168
Intermittents, 37
Intermodulation distortion, 14, 217
Isolating hum, 36
Isolating intermittents, 38
Isolating noise, 36
Isolation, stage by stage, 22

L

Lead dress, FM receiver, 55
Leakage, capacitor, 32
cathode-heater, 36
electrolytics, 33
IF transformer, 35
printed circuit, 35
Left-channel output, 72
Lift, tone arm, 202
Limiter, 50
Limiting, 57
Linear meter scale, 14
Line matching, 231
Line radiator speaker, 230
Lissajous figures, 28
Listening room, 211
L minus R channel, 61

L minus R signal, 72
Logarithmic meter scale, 14
Loss of RF gain, 32
Loudness controls, 88
Low-frequency feedback, 94
Low-impedance amplifier in-
put, 221
L plus R channel, 62
L plus R signal, 73
Lubrication, record changer,
207
tape recorder, 148

M

Main cam, record changer,
191
Maintenance, mobile tape
player, 185
tape recorder, 146
Meter scale, AC voltmeter,
14
Microphones, 237
Misalignment, 34
Mixer, 28
Mixer checks, 30
Mobile tape player, 183
Motorboating, 36
FM receivers, 55
Motor-drag drive, 139
Motor, record changer, 189
Multivibrator, bias/erase,
142
Multiplex, 61
Multiplex alignment, 70
Mute gate, 68
Mute switch, 144
Muting switch, record changer,
195

N

Needles, phono, 208
Negative feedback, 91, 144
Negative power supply, 127
19-kHz pilot signal, 62
Noise, 35, 76
auto radio, 179
FM, 55
record-change cycle, 195
Noise amplifier, 79
Noise-cancelling microphone,
239
Noise clipping, 47
Noise immunity circuit, 79
Noise level meter, 18

Noise limiting, 48
Nonchalant intermittents, 37

O

Omnidirectional microphone,
239
Operating cycle, record
changer, 191
Oscillator, 27
bias/erase, 141
FM, 48, 52
Oscillator coil, 26
Oscillator coupling, 31
Oscillator tests, 28
Output transistor checks, 25
"Over-live," environment,
223
Overload capacitor test, 40
Overloading, 92

P

PA amplifier ratings, 220
PA equipment, 219
Paging/talkback system, 235
Permeability-tuning, 172
Phono needles, 208
Pilot carrier trap, 67
Positive feedback, 97
Power output, amplifier, 217
Power supply checks, 21
Power supply hum, 36
Power supply regulation, 131
Preamp, 85
Preamp equalizers, 87
Preventive maintenance,
mobile tape player, 185
tape recorder, 147
Public address, 219
Pushbutton auto radios, 172
Push-pull bias/erase oscillator,
143
Push-pull output, 99
Push-pull output checks, 35

Q

Q point, 95
"Quieting," FM receiver, 76

R

Radial PA horns, 230

Ratio detector, 47
Ratio detector alignment, 59
Reactance-tube, 48
Rear speaker, 182
Receiver sensitivity, FM, 76
Record changers, 189
Record drop, 194
Recording bias, 141
Record/playback head, 140
Record/playback head re-
    placement, 148
Record-size selector, auto-
    matic changer, 202
Reflex trumpet PA speaker,
    226
Regeneration, 36
    FM receiver, 55
Regenerative feedback, 91
Regulation, power supply, 131
Resistor noise, 35
Resistor-type spark plugs,
    179
Response, AC voltmeter, 14
    amplifiers, 216
    frequency, tape recorder, 145
    microphone, 237
    sine/square-wave generator,
    14
Reverb equipment, 179
Reverberation, 223
Reverse-biased diode, 49
Reverse-charged battery, 168
Reverse torque, 139
RF carrier intercom, 243
RF gain loss, 32
RF interference, 112
RF signal injection, 21
RF tracking, 34
Right-channel output, 72
Ripple, 36
Room characteristics, 212
Rosin solder joint, 35
R/P head replacement, 148

S

Sales-demonstration facility,
    210
SCA, 62
Scale, AC voltmeter, 14
SCA traps, 67
Schmitt trigger noise circuit,
    79
Schmitt trigger noise limiter,
    48

Screen bypass, 36
Sensitivity, AC voltmeter, 13
    FM receiver, 76
    FM tuner, 57
    microphone, 240
    PA amplifier, 221
Separation, channel, 73
Series-voltage regulator, 132
70.7v speaker line, 235
Shut-off lever, record changer,
    193
Signal injection, 21
Signal-to-noise ratio, FM re-
    ceiver, 77
Signal-seeking tuner, 173
Sine/square-wave generator,
    14
Single-ended amplifier, 100
67-kHz subcarrier, 62
Solder, cold or rosin joint, 35
Sound pressure level, 217
Spark plugs, suppressor, 179
Speakers, 215
    auto, 178, 184
Speaker line matching, 231
Speaker types, 225
Speed control, mobile tape
    player, 187
    turntable, 189
Spillover, 73
Split-voltage power supply, 129
Spring-belt drive, 139
Spurious oscillations, 36
Square-wave generator, 14
Squealing, 37
Stabilization, FM oscillator,
    48
Stabilizing resistor, 82
Stereo cartridge player, 164
Stereo detector, 67
Stereo generator, 75
Stereo headphones, 216
Stereo indicator, 69
Stereo multiplex alignment, 70
Stereo cartridge tape player,
    183
Stereo tuner, 67
"Storecasting," 62
Styli, 208
Stylus set-down, 206
Subsidiary Communications
    Authority, 62
Substitute speaker, 34
Substitution, capacitor, 33,
    56
    oscillator, 30

Substrate amplifier, 115
Suppressor spark plugs, 179
Suspension, turntable, 199
Switching, AM/FM combinations, 120
Switch lever, record changer, 191
Synchronous multiplex detector, 67

T

Tape cartridge, 157
Tape head adjustment, 153
Tape head cleaning, mobile, 185
Tape head current measurement, 155
Tape head degaussing, 151
Tape transport, 139
Temperature-compensating capacitors, 54
Temperature variation, amplifier, 84
Test equipment, 12
Test tapes, 153
Thermistor stabilizing circuit, 96
38-kHz local oscillator, 62
38-kHz subcarrier, 61
Tire static, 179
Tone arm, 199
Tone arm adjustment, 202
Tone arm height, 206
Tone arm positioning, 193
Tone control, 88
Torque measurement, 148
Total-harmonic analyzer, 16
Total harmonic content, 220
Tracking pressure, 206
  tone arm, 202
Transformer checks, 34
Transistor checks, 25, 31
Transistor replacement, 117
  auto radio, 178
Transport, cartridge tape, 159
  tape, 139

Trap, pilot carrier, 67
Traps, SCA, 67
Treble control, 88
Trumpet-type PA speaker, 226
Tuner sensitivity, 57
Turntable drive, record changer, 189
Turntable suspension, 199
TVM, 14

U

Unidirectional microphones, 239
Ultra-cardioid microphones, 239

V

Variable acoustic control, 215
Variable-capacitance diode, 49
Variable-voltage transformer, 40
Varicap AFC, 49
Vertical adjustment, 202
Vibration-sensitive intermittents, 37
Voltage divider, 33
Voltage-divider bias network, 82
Voltage feedback, 96
Voltmeter, AC, 13
Volume control, 88
VTVM, AC, 13

W

Wave analyzer, 14
Weak battery, 32
Weak receiver, 32
Wheel static, 179
Wiring, FM, 55
Wireless intercom, 243
Wow, 17
  tape recorder, 145

Z

Zener diode, 132